PREFACE TO THE LIMPBACK EDITION

and rookeries have retreated to as far out as Denham in Bucks, Theydon Bois in Essex, Sevenoaks in Kent and Walton Downs in Surrey. Several introduced species of bird have greatly increased (Canada goose, gadwall, mandarin duck) or arrived (ruddy duck, rose-ringed parakeet), while the collared dove and little ringed plover have come by natural means. Much larger numbers of cormorants are also to be seen in the London area at all times of year (although they do not breed) than in the 1940s.

One of the most notable changes has been the freeing of the tidal portion of the Thames from the grosser forms of pollution, which during the first two-thirds of the present century had made that part of the river which passes through the main metropolitan area virtually devoid of both fish and invertebrate life. Today such freshwater fish as dace, roach and bleak can be caught right in the middle of London, for instance off the steps of County Hall, and flounders once again ascend the river from the lower estuary. There are signs too that the reintroduction of salmon into the upper reaches of the river may yet lead to the Thames regaining some of its former reputation as one of the great salmon rivers of Europe. At the same time waterfowl such as shelduck, pochard and tufted duck, and waders such as lapwing, redshank and curlew, have been coming much further up-river from the lower estuary than they have done at any time in the past 150 years.

The most important change affecting plant life has been the disappearance of many, if not most, of the elms of the London area, following the serious and widespread Dutch elm disease epidemic of the 1970s. While many of the rarer plants have become still rarer, the London rocket, *Sisymbrium irio,* considered virtually extinct in 1945, has reappeared both at the Tower of London and the Zoo in Regent's Park. However, most of the plants which have increased in the London area in the past forty years are well known weeds, both brightly coloured ones such as Oxford ragwort and rosebay willow-herb and the more subfusc, such as Canadian fleabane and gallant soldier *Galinsoga.*

R.S.R.F.
Chinnor Hill, 1984

The aim of this series is to interest the general reader in the wild life of Britain by recapturing the inquiring spirit of the old naturalist. The Editors believe that the natural pride of the British public in the native fauna and flora, to which must be added concern for their conservation, are best fostered by maintaining a high standard of accuracy combined with clarity of exposition in presenting the results of modern scientific research.

THE NEW NATURALIST

LONDON'S
NATURAL HISTORY

by

R. S. R. FITTER

WITH 93 PHOTOGRAPHS
BY ERIC HOSKING AND OTHERS
AND 12 MAPS AND DIAGRAMS

COLLINS   ST. JAMES'S PLACE   LONDON

THE AUTHOR'S best thanks are due to the following, and to others too numerous to mention, for information, help, advice, criticism and forbearance, while he was writing this book : the Director of the British Mosquito Control Institute ; Mr. C. P. Castell ; Mr. R. Preston Donaldson, Secretary of the Royal Society for the Protection of Birds ; the Superintendent of the Epping Forest Conservators ; Mr. F. J. Epps, Hon. Secretary of the South-Eastern Union of Scientific Societies ; Dr. H. Godwin, F.R.S. ; Mr. M. A. C. Hinton, F.R.S. ; Mr. A. S. Kennard; Mr. R. E. Latham, for help with the translation of Appendix A from the Latin; the Clerk of the London County Council; all those members of the London Natural History Society whose assiduity in amassing records of the natural history of the London area has made this book possible ; Mr. J. E. Lousley ; Dr. M. T. Morgan, Medical Officer of Health of the Port of London ; the Museums Association ; the National Smoke Abatement Society ; the Editors of the *New Naturalist* ; Mr. E. R. Parrinder ; Professor C. E. Raven ; Professor James Ritchie ; Dr. E. J. Salisbury, F.R.S., Director of the Royal Botanic Gardens, Kew ; Messrs. G. W. Scott & Sons, basket manufacturers ; Mr. G. B. Stratton, Librarian of the Zoological Society of London ; Mr. G. H. Underhill, of the Metropolitan Water Board ; the Clerk of the Vintners' Company ; and the Librarian of Dr. Williams's Library. To my wife I am indebted for an especial degree of forbearance, as well as for help, advice, criticism and information.

R.S.R.F.

HIGHGATE,
October, 1945.

First published in 1945 by
Collins 14 St. James's Place London
Produced in conjunction with Adprint

CONTENTS

Acknowledgments   iv
Editors' Preface   xii

CHAPTER

1 Introduction   I

2 Before Londinium   8

3 Romano-British London   27

4 Medieval London   34

5 The Wen Begins to Swell   61

6 The Wen Bursts   101

7 Nature Indoors   111

8 The Wholly Built-up Areas   119

9 The Partly Built-up Areas : Gardens and Parks   134

10 The Effects of Digging for Building Materials   148

11 The Influence of Trade and Traffic   152

12 The Influence of Water Supply   162

13 The Influence of Refuse Disposal   171

14 The Influence of Smoke   179

15 The Influence of Food-Getting   185

16 The Influence of Sport   193

17 The Cult of Nature   207

18 The Influence of the War   228

CONTENTS

APPENDIX

A   A Botanical Ramble in 1629   244

B   The Birds of the London Area   249

C   The Lost Plants of Middlesex   254

D   The Hundred Commonest London Plants   256

E   Natural History Societies, Museums, etc.   259

F   List of Flowering Plants and Ferns Recorded from Bombed
     Sites in London.   By E. J. Salisbury, D.Sc., F.R.S.   265

Bibliography   269

Index   277

LIST OF COLOUR PLATES*

FACING PAGE

1   Beeches and hollies at Ken Wood   **4**

2   Flooded gravel pit near Harefield, Colne Valley   **5**

3   Docks and thistles on waste ground by flooded gravel pit in Colne Valley, near Harefield   20

4   Sand-martin colony in disused sand pit near Barnet By-pass, North Mimms   21

5   Pollard hornbeams at Epping Forest, near Loughton   24

6a   The river wall at the confluence of the Ingrebourne with the Thames at Rainham, Essex   25

6b   The Upper Pool at sunset : London Bridge in the background   25

7   Feeding the pigeons in front of St. Paul's Cathedral   32

8a   A rabbit warren in Hainault Forest, Essex   33

8b   A badger holt at Loughton Camp, Epping Forest   33

9   The New River and the North-east Reservoir at Stoke Newington   36

10a   Aldenham or Elstree Reservoir, Herts.   37

10b   Walthamstow, No. 5 Reservoir, showing the heronry island   37

11   The famous maidenhair or ginkgo tree at Kew Gardens   52

12   The nearest rookery to London, at Lee Green, S.E.12   53

13a   Allotments on Barking Level, Essex   68

13b   Backyard pig farming   68

14   Cock Blackbird at nest   69

15   Cabbage attacked by Caterpillars   84

16a   Red admiral and small white butterflies at Michaelmas daisies   85

16b   Comma butterfly on laurel   85

17   Chickweed, groundsel, field speedwell, red dead nettle and annual meadow grass on market garden land near Chelsfield, Kent   88

18   The duelling ground at Ken Wood   89

19   A magnolia in the grounds of Ken Wood House   96

20a   The Thames at Hammersmith with mute swans in the foreground   97

20b   Teddington Lock   97

21a   A black-headed gull being fed in St. James's Park   120

*As explained on the first page, these plates are here reproduced in black and white.

FACING PAGE

21b Waterfowl in St. James's Park—South African grey-headed sheld-duck, a pair of mallard and a coot   120

22 Sheep grazing in front of Ken Wood House   121

23a Coldharbour Farm, Mottingham—the last farm in London   128

23b A plum orchard near Chelsfield, Kent   128

24a Goldfinch on nest   129

24b A pair of mute swans at their nest by the River Lea near Hertingfordbury, Herts.   129

25 Anglers on the River Lea near Broxbourne, Herts.   132

26a A bold red deer at Richmond Park   133

26b Feeding the pelicans at St. James's Park   133

27 Hainault Forest, Essex, from Dog Kennel Hill. The whole of this area was ploughed up a hundred years ago   148

28a Daffodils at Ken Wood   149

28b Crocuses at Hyde Park Corner   149

29 Tulips in St. James's Park   164

30 Horse-chestnuts in Bushy Park   165

31a Almond blossom in suburban front gardens, Ruislip, Middlesex   180

31b Roses in Queen Mary's Garden, Regent's Park   180

32 Pear tree in blossom, Crouch End, Middlesex   181

33 Azaleas in Kew Gardens   196

34 Forsythia on Shepherd's Hill, Highgate, looking across to Alexandra Palace   197

35 Bluebells in Oxhey Woods, Herts.   212

36 Highgate Wood, showing the fenced-off bird sanctuary   213

37 Rosebay willow-herb and Canadian fleabane in a ruined City church   216

38 Coltsfoot on a blitzed site   217

39 Berkeley Square plane trees   224

40a Glasshouses in the Lea Valley   225

40b Cress-beds at Fetcham, Surrey   225

LIST OF PLATES IN BLACK AND WHITE

FACING PAGE

I   Aerial view of London and the Thames Estuary   40

II   Central London, from a tall building near St. Paul's   41

III   The great vine at Hampton Court   48

IV   The Regent's Canal as it passes through the Zoological Gardens   49

V   A horse-chestnut in bloom by the Grand Union Canal in Cassiobury Park, Watford   56

VIa   A brown rat in a London warehouse   57

VIb   Nests of house-mice in flower-pots in a greenhouse   57

VIIa   Male and female cockroaches with egg-sac   64

VIIb   Spiders' webs   64

VIII   A cedar tree in Highgate cemetery   65

IX   The Chelsea Physic Garden   72

Xa   A typical Inner London street with no gardens   73

Xb   Garden of a blitzed suburban house overgrown with rosebay willow herb and other weeds   73

XIa   Cock black redstart leaving nest-hole in the Temple   80

XIb   Hen black redstart with food for young   80

XII   Starlings roosting in plane trees at Marble Arch   81

XIII   Typical suburban back gardens   104

XIVa   Unusual site of blackbird's nest   105

XIVb   Blue tit at coconut   105

XVa   House-sparrows feeding from the hand in a London park   112

XVb   A grey squirrel in a London park   112

XVIa   Blue tit carrying food to nest in lamp-post   113

XVIb   Woodpigeon on nest in plane tree near the Bank of England   113

XVII   Robin's nest in old watering-can   152

XVIIIa   Black fly on broad bean   153

XVIIIb   An aspidistra in a London parlour   153

LIST OF PLATES IN BLACK AND WHITE

FACING PAGE

XIX   The Upper Pool from the south end of Tower Bridge   160

XX   Chiswick Eyot from Chiswick Mall   161

XXI   Feeding black-headed gulls on Victoria Embankment with old Waterloo Bridge in the background   168

XXII   Muswell Hill Golf Course with Alexandra Palace in the background   169

XXIII   A typical suburban playing field   176

XXIVa   A melanic form of the Peppered Moth   177

XXIVb   A London stray cat   177

XXV   Waste ground at Bromley-by-Bow gasworks   184

XXVI   A typical backyard poultry run   185

XXVII   Haymaking in Green Park   192

XXVIII   Anglers at Highgate No. 3 pond with Ken Wood in the background   193

XXIX   A giraffe at the Zoological Gardens in Regent's Park   200

XXX   Little owl with cockchafer at nest-hole   201

XXXIa   Cormorants in St. James's Park   208

XXXIb   Two mallard on planks floating in a static water tank in London   208

XXXII   Oxford ragwort on a blitzed site near St. Paul's   209

Every care has been taken by the Editors to ensure the scientific accuracy of factual statements in these volumes, but the sole responsibility for the interpretation of facts rests with the Authors.

DIAGRAMS AND MAPS

DIAGRAM

1 Section through the London Syncline   10

2 Section through the Thames gravel terraces   10

MAP

1 The Physical Features of the London Area   12-13

2 The London Area in Roman Times   29

3 The Spread of London, 1560-1836   66-67

4 The Spread of London, 1872-1935   102-103

5 The City of London, showing important streets and buildings   117

6 The West End, showing important streets and buildings   124-125

7 Hampstead Heath and Highgate Woods   145

8 The County of London, showing the Metropolitan Boroughs   156-157

9 Epping Forest   223

10 The Greater London Area, North of the Thames   240-241

11 The Greater London Area, South of the Thames   242-243

EDITORS' PREFACE

R. S. R. FITTER is a young social scientist and writer who has been a naturalist all his life. Though he has never been in the strict sense a professional biologist, the study of animals and plants has occupied so much of his leisure that he can by no means be described as an amateur. Indeed, he holds two positions which, though honorary, are as onerous as they are important—the Secretaryship of the British Trust for Ornithology and the Editorship of *The London Naturalist*, the journal of the London Natural History Society.

Mr. Fitter has always lived in London, as have his father, grandfather and great-grandfather ; and he has made a special study of London's natural history—and the history of its natural history—for over ten years. He has, clearly, the material qualifications for the work he has chosen to do ; and the reader will soon agree that he has done it well. And it is time that it was done—high time that this book was written. For up to now there has been no real attempt, in any biological literature we are familiar with, to write the history of a great human community, in terms of the animals and plants it has displaced, changed, moved and removed, introduced, dispersed, conserved, lost or forgotten. In certain ways Mr. Fitter's book makes gloomy reading, for the progressive biological sterilisation of London is a sad history. But the discerning reader will soon notice that the sterilisation is not complete. Indeed, in this remarkable history not all is on the debit side. There is the fascinating story of the adaptation of wild life to an environment which is almost wholly man-made. There is also the fact that London natural history to-day has its special compensations, even its new and particular treasures.

Mr. Fitter's book is a notable contribution to the history and the understanding of the processes at work in the evolution of London's wild life, and will be of help in the framing of any future policy.

THE EDITORS

INTRODUCTION

LONDON is the largest aggregation of human beings ever recorded in the history of the world as living in a single community. In 1931, the date of the last census, not far short of nine million men, women and children were living within twenty miles of St. Paul's Cathedral. In no other area of equal size in the world are as many as 7325 people to the square mile to be found.

Such a huge mass of people cannot settle on the soil of a district without causing devastating changes in the natural communities of animals and plants that lived there before the men came. It is the main aim of this book to trace the story of these changes, to show the influence of man as a biotic factor in the most extreme example of urban development ever known. To do this it is first necessary to go back to the time when there was no human settlement on the present site of London. The spread of London, at first gradual, but latterly, especially in the past hundred years, relatively very rapid, is then chronicled. Finally, the present balance between man and the natural communities, showing the degree of adaptation which the latter have achieved, is described.

Up to the invasion of the Romans in A.D. 43 there is no certain evidence of the existence of any permanent settlement on the famous square mile of the City, which has been the core of London for the past nineteen hundred years. Vast changes had taken place in the flora and fauna of the lower Thames valley since the first men had come there some hundreds of thousands of years before, but they were due far more to long-term climatic trends than to any intervention of the men of the Stone, Bronze or Iron Ages. Man still fitted into an ecological niche among the other communities ; he had not yet come to dominate them all.

With the coming of the Romans the first city was built on the twin hillocks on either side of the Walbrook rivulet, and man became a decisive influence within the wall that was built to enclose and protect Londinium. When the Romans left, and before the Saxons finally took possession, Nature must have reconquered much of her lost ground. From the seventh to the eleventh centuries there was much open ground within the City walls, but thereafter London not only filled out the space within its own walls, but overspilled, and began a rake's progress that is not ended yet, with almost continuous tongues of built-up area stretching from Hertford in the north to Reigate in the south, and from Tilbury in the east to Slough in the west.

Not the least interesting part of the story is the high degree of adaptation which the animals and plants of the lower Thames valley have shown to the immense changes wrought by man. Even where human activity has created wholly artificial habitats, a flora and fauna have in the course of centuries adapted themselves to conditions some-times totally unlike anything normally found in Nature. The fauna of houses, and especially of warehouses, is extraordinarily varied and numerous. Three mammals, the black rat, the brown rat and the house-mouse, have succeeded in adapting themselves to a completely indoor existence, and a large number of invertebrates, spiders, flies, lice, bugs, clothes-moths, cockroaches, and so forth have done the same with varying degrees of success.

Out of doors, but still in areas completely or almost completely covered by roads, railways and buildings, three or four species of birds can live comfortably. This is true notably of the house-sparrow and the London pigeon, while the black-headed gull is developing a habitat-preference for railway sidings in the London area. Many of the lowlier plants, such as the mosses, also contrive to exist in habitats which simulate for them a rocky cliff or hillside, and wherever a bare patch of ground appears in the heart of the built-up area, a host of flowering plants, like the rose-bay willow-herb, Oxford ragwort, colts-foot and Canadian fleabane,[1] spring up in a remarkably short time.

Where the built-up areas are partially diluted with gardens, parks and other open spaces, several species of birds and innumerable insects and other invertebrates have carved out niches for themselves. The garden association of birds is now as definite an avifaunal community as that associated with heathland or sea-cliffs. It comprises the starling, greenfinch, chaffinch, house-sparrow, great tit, blue tit, mistle-thrush, song-thrush, blackbird, robin, hedge-sparrow, wren, house-martin and

[1]*Epilobium angustifolium ; Senecio squalidus ; Tussilago farfara ; Erigeron canadensis.*

swift as permanent members, with other species, such as the linnet, pied wagtail, spotted flycatcher, willow-warbler and tawny owl often added. All these birds originally had quite different habitats, in woods or scrubland, or (in the case of the house-martin) on cliffs, but have quite adapted themselves to life in suburbia.

In the flower-beds, and wherever else man has broken the ground to make an artificial seed-bed, miniature forests of seedling plants arise. The weeds of cultivated ground are a community peculiarly dependent on human influence, and grow luxuriantly in a type of habitat which before man began to till the soil must have existed in the London area only as a result of rare accidents. They include such familiar pests of the gardener as the charlock, shepherd's purse, chickweed, creeping cinquefoil, scentless mayweed, groundsel, sow-thistle, dandelion, field bindweed, speedwell, red dead nettle, greater plantain, white goose-foot or fat-hen, garden orache, broad dock, black bindweed, sun spurge and annual meadow grass.[1] Some of these weeds are aliens that have followed man to most parts of the world where he has stayed to cultivate. Others have originated locally.

Many facets of man's social activities in and around London have left their mark on the animal and plant communities, destroying some, creating others. When birds like the rook and jackdaw have been driven ever farther from St. Paul's by the relentless advance of London's tide of bricks, others like the wood-pigeon and moorhen have moved in to colonise the desirable new habitats created by the isolation of the parks as oases of greenery. The influence of international trade has brought many new creatures to the port of London, most of them undesirable, such as the rats and cockroaches. The construction of docks to hold the ships that engage in this trade has made at the same time vast artificial fish-ponds. The canals that lead to those docks have preserved strips of greenery and a habitat for aquatic birds such as the mallard in some very unlikely places, while the reservoirs that serve those canals form some of the most valued sanctuaries of wild life in the London area. The roads and railways which carry the trade and traffic of London, besides sterilising large areas of land, levy a toll on the remaining wild life of the district.

The arrangements man has made for supplying himself with water

[1] *Brassica sinapis ; Capsella bursa-pastoris ; Stellaria media ; Potentilla reptans ; Matricaria inodora ; Senecio vulgaris ; Sonchus oleraceus ; Taraxacum officinale ; Convolvulus arvensis ; Veronica agrestis ; Lamium purpureum ; Plantago major ; Chenopodium album ; Atriplex patula ; Rumex obtusifolius ; Polygonum convolvulus ; Euphorbia helioscopia ; Poa annua.*

exercise a very powerful influence over the aquatic animal and plant communities of the London area. In addition to the algæ, sponges, molluscs and other creatures that may inhabit waterworks and water-pipes, many kinds of birds and fishes live in or visit the great reservoirs in the Thames and Lea valleys. Practically every duck, grebe and diver that visits the British Isles regularly can be seen on one or other of the London reservoirs within the space of two or three years.

In the Middle Ages several creatures thrived in London as scavengers, notable the raven and the kite. Later on, the Thames became so fouled with sewage that it lost almost all the fish in the Central London reaches. More recently the development of sewage farms has created both an interesting fauna and flora of flies, algæ and other creatures, and such a good replica of the primeval Thames marshes that thousands of wading birds halt on migration in the London area each year. The newest scavengers of London are the gulls, who in the short space of fifty years have become almost the typical birds of London.

The influence of the smoky atmosphere of London on the flora and fauna has been far-reaching. London saved its woodlands by importing " sea-coal " from Newcastle, only to find that the tarry deposits that resulted from burning thousands of tons of raw coal every year effectively inhibited the growth of many plants, and so drove out also the insects that fed on the plants and the birds that fed on the insects.

Around the fringe of the built-up area of London there has always been a fringe of cultivation, agricultural and horticultural. The brick-earth soils of Middlesex and South Essex form one of the finest media in England for the cultivation of fruit and vegetables, and much of suburban London now covers with barren bricks and tarmac some of the richest market-gardening land in the country. The flora and fauna of cultivated land are a study in themselves, and from our point of view the most important thing is to show how they have been steadily driven outwards by the spread of London.

At one time the utilisation of wild animals and plants for food was not the least important influence of man on the wild life of the London area, but nowadays even rabbits and fish are shot and hooked more for sport than for the pot, and if pigeons are trapped in Trafalgar Square from time to time, it is because their numbers make them a public nuisance rather than because somebody wants pigeon pie.

Sport has always provided a powerful motive for man's interference with the balance of nature. Certain animals have been preserved so

PLATE I

ERIC HOSKING

Beeches and hollies at Ken Wood

PLATE 2

C. W. NEWBERRY

Flooded gravel pit near Harefield, Colne Valley

that they might be hunted, for instance, the deer, fox, partridge and pheasant, and to prevent their suffering the fate of earlier beasts of the chase, such as the wild boar. Other creatures have been persecuted because it was feared they might harm the game. Polecats, martens, owls and hawks have all been in this category from time to time. More modern manifestations of sport have resulted in large areas of open space being preserved in the form of golf courses and playing fields. These have been the means of enabling many animal and plant species to survive nearer the metropolis than would otherwise have been the case.

An increasing appreciation of the amenity value of wild animals and plants has had an important effect in the preservation of wild life in the neighbourhood of London. Large areas of open space have become sanctuaries for animals and plants. The value of such an open space as Hampstead Heath, for instance, in preserving many creatures that would otherwise have been driven much farther from the· centre of London can hardly be exaggerated. In many cases the desire of men to see animals and plants at close quarters has led to aliens being introduced, and first kept in captivity and then allowed to escape. The grey squirrel, little owl and Canada goose have been added to the fauna of the London area in this way, while many plants have escaped from cultivation and established themselves.

Both the 1914 and the 1939 wars have had important, but mainly temporary, influences on London's flora and fauna. The outstanding effect of the 1939 war is the creation of large areas of waste ground in the centre of the built-up area, which many plants, notably the rose-bay willow-herb, have rapidly colonised.

By and large, it is not too much to say that there is no blade of grass, no newt or mite or alga, within twenty miles of the centre of London that has not in some way had its way of life affected by the dominant species, man. Even more decisively than elsewhere in the British Isles, man has altered the distribution of natural habitats in the London area, has created some entirely new ones, and has virtually extinguished others. There is no fenland in the London area now, for instance, and bogs and swamps have almost gone. In 1912, for example, only five species, the water violet (*Hottonia palustris*), creeping willow (*Salix repens*), oval sedge (*Carex leporina*), moor-grass (*Molinia cærulea*) and mat-grass (*Nardus stricta*), remained out of the twenty-eight formerly recorded from the small bog near the Leg of Mutton Pond on Hampstead Heath.

WHAT IS THE LONDON AREA?

It is not easy to decide the area which should be covered in writing of the natural history of London. Most of the administrative boundaries are ludicrously inadequate. The County of London boundary is a line drawn arbitrarily on the map; it passes down the middle of Highgate High Street, for instance, and right through Ken Wood; where it divides Paddington from Willesden it passes through a maze of mean streets in which the format of the street name-plate is the only means of telling whether you are in the County of London or of Middlesex; the metropolitan borough of Woolwich has two curious little enclaves in the county borough of East Ham on the north bank of the Thames. The present area of the County of London is based ultimately on the area covered by the Bills of Mortality in the seventeenth and eighteenth centuries, and has about as much validity as a faunistic or botanical area as that enclosed by the Inner Circle railway. If an artificial boundary is to be chosen, it may as well be one that gives full scope for demonstrating the widespread nature of London's influence, which makes it necessary to go out for ten or fifteen miles in order to find many of the common habitats of the English country-side, such as woodland, heathland, or river marshes. The area covered by this book, therefore, is that within a radius of twenty miles from St. Paul's Cathedral, which happens to be the area covered by the London Natural History Society, the most active local natural history society in the London area.

London has always presented difficulties to British naturalists, because they like to record their species under the old geographical counties, and London sprawls over four or five of them. To-day London is connected by unbroken strips of wholly and partly built-up land with Middlesex, Surrey, Kent, Essex, and even with Barnet and Cheshunt in Hertfordshire. What is the sense of stopping before you reach the tram terminus, or counting a bird twice because it flies across the river? It is high time that naturalists reorganised their system to take account of real ecological dividing lines instead of the archaic boundaries of the Heptarchy, based on the number of men that could be raised to defend a Saxon nobleman's estates. Many of the existing boundaries were fixed for quite trivial reasons, and have lasted more than a thousand years. The present boundary between

Middlesex and Hertfordshire, for instance, was fixed so that all the lands of the Abbot of St. Albans might be in the same county—no doubt a laudable aim in the tenth century—with the result that the anomalous tongue of Hertfordshire almost engulfed in Middlesex between Finchley and Potters Bar owes its origin to administrative convenience in the reign of Athelstan.

The radius of twenty miles from St. Paul's Cathedral, taken as the boundary of the London area in this book, passes through the following places, going clockwise from the Thames estuary : Northfleet, Hartley, Ash, Kingsdown, Kemsing, Sevenoaks, St. Johns, Brasted Chart and Crockhamhill Common, in Kent ; Itchingwood Common, South Godstone, South Nutfield, Earlswood, Mead Vale, Reigate Heath, Betchworth, Box Hill, Fetcham Downs, Great Bookham, Banks's Common, Ockham Common, Byfleet, New Haw, Addlestone, St. Ann's Hill, Thorpe, Egham and Runnymede in Surrey ; Wraysbury, Sunnymeads, Datchet Common, Ditton Park, Wexham, Fulmer, Gerrards Cross and Chalfont Park in Bucks. ; Heronsgate, Chorley-wood, Micklefield Green, King's Langley, Bedmond, Prae Wood, Symondshyde Great Wood, Lemsford, Welwyn Garden City, Pans-hanger Park, Bengeo, Easneye and Eastwick in Hertfordshire ; and Little Parndon, Latton, Bobbingworth, Stondon Massey, Dodding-hurst, Shenfield, Thorndon Park, Bulphan, Orsett and Tilbury in Essex. (See Maps 10 and 11.) Longfield, Riverhead and Westerham in Kent ; Oxted, Redhill, Reigate, Leatherhead, Weybridge and Chertsey in Surrey ; Staines in Middlesex ; Colnbrook in Bucks. ; Rickmans-worth, St. Albans, Hatfield, Hertford and Hoddesdon in Herts. ; and Chipping Ongar, Brentwood and Grays Thurrock in Essex are just inside the boundary. Gravesend and Sevenoaks in Kent ; Dorking, Wisley and Virginia Water in Surrey ; Windsor in Berks. ; Slough, Stoke Poges, the Chalfonts and Chenies in Bucks. ; Sarratt, Hemel Hempstead, Harpenden, Wheathampstead and Ware in Herts. ; and Harlow, High Ongar, Hutton and West Tilbury in Essex are just outside the boundary.

BEFORE LONDINIUM

IN THE year 1877, at Meux's Horseshoe Brewery at the southern end of Tottenham Court Road, a well was bored, which plunged through 1146 feet of solid rock and millions of years of London history. If we reverse the process of drilling, and come up from the bottom of the well, we find ourselves first on a platform of Palæozoic rocks, laid down several hundred million years ago, that underlies the whole London region. This boring at Meux's brewery is now chiefly remembered as the first in the London area to reach down to the Devonian rocks of this Palæozoic platform.

Rising towards the surface, we pass through successive groups of rocks. Their sequence resembles what we should find if we travelled across the centre of England from the north-west to the south-east, though their thickness under London is much less than in most places where they are found at the surface, and many formations are absent. Leaving the Devonian platform, we pass first through 64 feet of the rocks of the Jurassic formation, similar to those which stretch across England from Dorset through the Cotswolds to North Yorkshire. Then follow 160 feet of Gault and 28 feet of the Upper Greensand, succeeded in turn by a great thickness, 655 feet, of chalk. A like succession of rocks is found in the Midlands as you approach the chalk escarpment of the Chilterns from the oolite ridge in Northamptonshire.

In the Cretaceous Period, tens of millions of years after the laying down of the Devonian platform, but still at so remote a distance from the present day that imagination boggles at it, the site of modern London lay below the surface of a clear, tranquil sea that stretched away northwards to the Pennines and westwards to the Welsh marches. In this sea, more than fifty million years ago, countless billions of minute creatures died every day for at least ten million years. Their hard shells sank into the chemically-deposited ooze at the bottom of the sea, and in the course of time the whole consolidated into the thick beds of chalk which form the cup of the London basin, its rim emerging to the north as the Chiltern Hills of Hertfordshire and Buckinghamshire and to the south as the North Downs of Surrey and Kent. (Map 1.)

The chalk is the first geological formation we have come to that still has an important influence on the landscape of the London area. (See Fig. 1.) North of the Thames it covers a large part of Hertfordshire north of the Colne and Lea valleys, with a number of outcrops on the south side of these two rivers, of which the most important is at Harefield in the extreme north-west corner of Middlesex. In southern Essex there is a small inlier from the Kentish chalk, and it is not generally realised that the steep hill on which Windsor Castle stands is an isolated outcrop of chalk. South of the Thames the chalk comes within a dozen miles of London at Croydon, and at Greenwich Park and a few other small outcrops comes appreciably nearer the centre.

The chalk has given Londoners two superb pieces of walking country in the Chilterns and the North Downs, both of which, being largely covered with a deposit of clay-with-flints, are more heavily wooded than the " blunt, bow-headed, whale-backed " South Downs. Epsom Downs on the Surrey chalk is the scene of the most popular of all London's outdoor occasions, the Derby, and Box Hill only a few miles to the south-west rivals Epping Forest on the north side as a resort of hikers and picnickers.

Few who have visited the scarp of the North Downs at Box Hill or above Oxted a few miles further east and admired the magnificent view across the Weald to the South Downs, can have failed to notice much nearer at hand great white gashes in the steep green slopes. These are the chalk-pits, which are also to be found widely dispersed and less conspicuously all over the chalk, for instance at Riddlesdown in Surrey, Harefield in Middlesex, and Grays in Essex. The chalk near London has been worked by man for many centuries, first for agricultural and later also for industrial purposes. As late as 1909 small quantities of chalk were being quarried in the County of London, at Charlton.

Many fossils of sea-urchins and other echinoderm inhabitants of the great Cretaceous sea can be found in the chalk of the London area ; for instance, *Micraster cor-anguinum*, *Echinocorys scutatus* and *Conulus albogalerus* in the Woolwich district.

The chalk was raised to the surface of the sea by vast movements

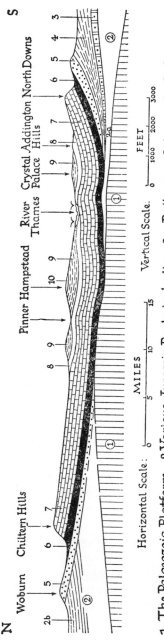

N

Woburn Chiltern Hills Pinner Hampstead River Crystal Addington North Downs S
                                          Thames Palace Hills

Horizontal Scale:

MILES
0   5   10   15   20

Vertical Scale.

FEET
0   1000   2000   3000

1, The Palaeozoic Platform 2, Various Jurassic Rocks including 2a, Bathonian Oolites 2b, Oxford Clay
3, Purbeck Beds 4, Wealden 5, Lower Greensand 6, Gault 7, Chalk
8, Lower London Tertiaries (Landenian) 9, London Clay 10, Bagshot Sands ⌇Alluvium

DIAGRAM 1 Section through the London Syncline.
Note that the places mentioned are not necessarily in a straight line.
(Redrawn from "An Introduction to Stratigraphy" by L. Dudley Stamp, 1934).

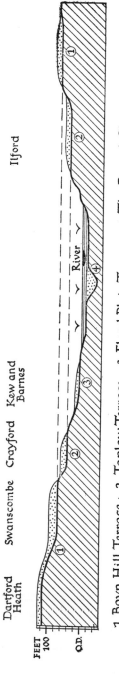

Dartford Swanscombe Crayford Kew and Ilford
Heath   Barnes

FEET
100

O.D.

1, Boyn Hill Terrace 2, Taplow Terrace 3, Flood Plain Terrace 4, The Buried Channel ⌇Alluvium.

DIAGRAM 2 Section through the Thames gravel terraces.
Note that the places mentioned are not necessarily in a straight line.
(Redrawn from "An Introduction to Stratigraphy" by L. Dudley Stamp, 1934).

of the earth, and then soon after (geologically speaking) lowered again.
On top of it were laid down a series of sands and clays wi h some p bble-
beds, belonging to the period known to geologists as the Eoccne. These
are represented in the Meux's well-boring by 21 feet of Thanet Sands
(much less than is found in the famous pits at the surface at Charlton)
and 52 feet of Woolwich and Reading Beds, which are mixed sands,
clays and pebble-beds laid down in shallow, partly brackish water.
These lower Eocene deposits occur on the surface mainly in South-east
London and the adjacent parts of Kent, where they often form a rich
loamy soil of considerable agricultural and horticul.ural importance.
The beds immediately above the Woolwich and Reading series are
known as the Blackheath Pebble Beds, taking their name from the
bleak plateau which they help to form. The Blackheath beds yield
a barrener soil than the other Lower London Tertiaries, and are
responsible for the survival of several picturesque open spaces in
South-east London and North-west Kent, such as Plumstead, Chislehurst, Keston and Hayes Commons and Bostall Heath, where birches,
Scots pines, gorse, broom and heather thrive.

Above the Woolwich and Reading Beds in the Meux's boring
comes the great Eocene deposit of the London Clay, only 64 feet thick
here in Tottenham Court Road, but anything up to 400 feet or even
more in other parts of the London basin. The London Clay was laid
down in a sea whose shores were covered with a vegetation of extinct
plants, which most closely resemble those now found in subtropical
lands. It is the most important of all the rocks that appear above the
surface in the London basin, as it covers large parts of Middlesex,
South Hertfordshire, South-west Essex and North Surrey, and pro-
duces a landscape of smooth, flat undulations, with elm-lined hedgerows as a typical feature. Its stiffness and intractability evokes bad
language from many suburban gardeners, who can console themselves
for aching backs by the reflection that it is an excellent medium for
the cultivation of roses.

A long period of geological time elapsed after the deposition of the
London Clay, during which great earth-building movements took
place in other parts of the world. The Himalayas and the Alps arose,
but the main effect of these movements on the London area was to
fold the chalk and the overlying beds into a gentle basin—the London
Basin. The chalk is a water-bearing formation, and water trapped
between it and the clays can be reached under London in artesian

MAP 1   The Physical Features of the London Area.
(Based on the ½″ Ordnance Survey
Map of Greater London, 1935)
By permission of H.M. Stationery Office

RESERVOIRS

1 Aldenham
2 Barn Elms
3 Brent
4 Island Barn
5 King George V
6 Kempton
7 Hampton
8 Lonsdale Road
9 Queen Mary
10 Ruislip
11 Staines
12 Stoke Newington
13 Walthamstow
14 West Molesey

---- CANALS AND
      LEA NAVIGATION

HEIGHTS IN FEET

0 – 100

100 – 200

200 – 300

300 – 500

OVER 500

SCALE OF MILES

0   2   4   6   8   10

wells. This abundant water-supply under London's doorstep helped in the early development of London as an industrial centre.

At the end of the London Clay period a series of sandy deposits, sometimes mixed with clay and pebbles, were laid down in the waters of a shallow sea. These were the Claygate, Bagshot, Bracklesham and Barton Beds, none of which are represented in the Meux's boring. The Bagshots, however, are important elsewhere in the London area, particularly at Hampstead and Highgate, where the twin hills that overlook North London have caps of Bagshot Sand on the London Clay. The dryness of the soil, and the plentiful springs at the junction of the sand with the clay, have combined to make these two villages, now suburbs, noted health resorts for Londoners for many years. In the eighteenth century the chalybeate spring, at present commemorated in Well Walk, was a popular centre of public entertainment, and a small enclave of Hampstead on the Heath itself is known as the Vale of Health. There is still a small chalybeate spring on the West Heath, near the Leg of Mutton Pond. The sands of Hampstead Heath have long been dug by the Lords of the Manor and others, and so much of the subsoil has been removed that the Spaniards Road which divides the East from the West Heath has the appearance of a causeway. As late as 1939 the sandpits of the East Heath were extensively dug to provide sand for the sand-bags at the outbreak of war. A classic exposure of the Bagshot Sand may still be seen about a hundred yards on the Highgate side of the Spaniards Inn, by looking over the fence that divides the main road from Ken Wood. In this pit, which contains seams of clay or loam, with iron sandstone, and a pebble bed near the base, sand-martins nested as late as 1926.

With the last 22 feet of the Tottenham Court Road boring, consisting of gravel, alluvium and topsoil, we at length reach a horizon where man himself begins to impinge on the scene of so many of his later triumphs and infamies. After a further gradual raising of the land above the sea, came a depression, which produced something that a modern Londoner would begin to recognise as the view he gets looking north from Epsom Downs or south from Highgate Hill. (Plate I.) Somewhere about a million years ago a great river, the forerunner of the modern Thames, flowed eastwards between the two chalk slopes to join the forerunner of the modern Rhine somewhere in the southern part of what is now the North Sea. Already, some 60 miles to the southward, there must have dwelt a tribe of men (*Eoan-*

thropus), one of whose skulls was dug up at Piltdown in 1912, in association with the remains of elephants, mastodons, hippopotami, rhinoceroses and sabre-toothed tigers. Whether any Piltdown men lived in the valley of the early post-Pliocene Thames we have no means of knowing, though it seems quite probable. We do know, however, that the lower part of the Thames valley was inhabited by the men of the Old Stone Age, who appeared there at least 500,000 years ago.

Between the probable presence of the Piltdown men of the Eolithic culture on the banks of the Thames and the known presence of men of the Chellean and Acheulean Palæolithic cultures there began the violent series of climatic changes known as the Ice Age. Vast glaciers and ice-sheets spread at different times and from different directions over practically the whole of England north of the Thames. Once a tongue of ice came as far as Finchley, bringing with it boulders of rock from far to the north, as well as clay and finely crushed rock from the lands over which it had travelled. When the ice melted, the boulders embedded in clay were left behind in the deposit we know as boulder clay. So it is that we find large lumps of chalk from Lincolnshire in the tongue of boulder clay that to-day stretches from Whetstone to East Finchley Station. It seems probable that this glacier did not follow the present line of the Great North Road any further, but divided against the heights of Hampstead and Highgate, sending one arm to Hendon. Streams of melt-water from these Pleistocene glaciers deposited coarse gravels and sands, often closely associated with the boulder-clay, but giving rise to porous, sterile soils. There are other high-level gravels in the London area, of which some may have been deposited by the precursor of the Thames, and some may be even older. These, like the glacial gravels, afford poor soils, so that the land was not worth enclosing in Tudor and Georgian times. Modern London thus owes many of its principal lungs, such as Wimbledon Common, to these geological accidents of hundreds of thousands of years ago. By the time the rising building values made them worth enclosing, a vigilant public opinion was able to thwart the attempt.

Gradually the Thames cut its bed down to lower levels, while the level of the land rose. Terraces of gravel have been left along its former courses to mark the different stages in its history. (Fig. 2.) Three main terraces can be distinguished. The High or Boyn Hill Terrace, laid down probably some 100,000 years ago, is the oldest, and is now generally 100-120 feet above the present level of the river. The Middle or Taplow Terrace is now about 50 feet above the river, the Low or Flood Plain Terrace was probably contemporaneous with the end of the last glacial period some 10-15,000 years ago, and is only about 25 feet above the present river level. These gravel deposits contain numerous remains of early man as well as of the animals and some-times of the plants that lived at the same time as these first Londoners. It must not be supposed, however, that the Thames ever spread right across the valley from terrace to terrace at any one time. These gravels rather mark the northern and southern limits of the meandering of a river that was both much broader and much swampier than anything we have known in historic times. Similar terraces were deposited in the neighbouring valley of the Lea, the most important tributary of the Thames in the London basin.

When the Boyn Hill gravels were being laid down by the river water, part of the Ice Age was past, though Arctic conditions would come again more than once. Men of the Chellean culture of the Palæolithic or Old Stone Age were already there, for we have found their stone implements embedded in the gravel. They lived in the relatively warm climate of one of the interglacial periods, on the banks of a wide river, with many eyots and backwaters, and bordered by a dense forest. They hunted, or were hunted by, the many other mammals inhabiting the forest and the river. These included the mammoth (*Elephas primigenius*), straight-tusked elephant (*Elephas antiquus*), the great wild ox known as urus (*Bos primigenius*), wild horses (*Equus caballus*), red deer (*Cervus elaphus*), reindeer (*Rangifer tarandus*), narrow-nosed rhinoceroses (*Rhinoceros leptorhinus*), wild boars (*Sus scrofa*), wolves (*Canis lupus*), giant beavers (*Trogontherium cuvieri*), and also voles (*Microtus*) and field-mice (*Apodemus sylvaticus*). The presence of both the sub-tropical straight-tusked elephant and the sub-arctic mammoth in the same deposits is suggestive of the great changes of climate that were taking place at this time. As the glaciers swept southwards they gradually drove the warmth-loving animals south-wards before them ; as the ice retreated northwards again, the warmth-loving animals reconquered their lost ground, driving before them the great shaggy beasts that had come down from the north with the ice.

Man at this time can have had little effect on his natural environ-ment, either plant or animal. We presume that he was an eater of meat and berries, but there is no reason to suppose that he upset the balance of Nature to any appreciable extent. Though he was already less firmly wedged into a niche than almost any other living thing, he still ate some and was eaten by others. Only when a creature ceases to be a regular item on some other creature's menu can it be considered to be well on the way to dominance. In two respects only was man out and away above the rest of the mammals in the competitive race : he had already discovered the use of fire and of tools. We may suppose that accidental forest fires were among the first large-scale impacts of man on his natural environment.

Attempts to reconstruct the life of Chellean man from such meagre circumstantial evidence as we have usually turn out to be literary rather than scientific exercises. However, H. R. Hall's description of what Chellean man himself looked like is of some interest :

" He was shorter than the average Englishman of to-day. He stood with his knees bent, his body stooping, and his head low in front of his hunched-up shoulders. . . . His legs and arms, chest and shoulders, were probably covered with hair growing so thick that it was almost like fur. . . . His nose was broad and flat, with wide nostrils to enable him to hunt by scenting out his prey. His ears were longer and larger than ours, helping him to catch every sound—a movement in the bushes, the snapping of a twig. . . . He had bushy eyebrows sprouting above his low brows, but the most strange feature of his face was the lower jaw. We should say that he had no chin, for his jaw sloped away into his thick neck. He showed his big teeth with every movement of his hairy lips, and altogether he was not at all the sort of man we should care to meet in a lonely spot."

The first men were probably vegetarians only, eating berries, roots and leaves, but as the ice came southwards and the weather grew steadily colder they were driven to a diet of flesh. Not only did man have to catch and kill the mammals and birds and fish whose flesh he ate, but he had to compete for his prey with other mammals, such as lions and bears and wolves, which had been longer on the job, and were just as ready to dine off man as off deer. It was probably in the course of this intense competition in a bitterly cold climate that the great discovery of the use of flint for implements was made.

With this discovery began the culture which we call the Stone

Age, which lasted in Britain from about 500,000 years B.C. almost up to the era when written history begins ; the final phase, the New Stone Age, overlapped with the Early Bronze Age about 2000-1750 B.C. As generations of Chellean men gradually became more skilled in the manipulation of flint, over a period of some 200,000 years, the Chellean culture passed into what we now know as the Acheulean, which itself lasted for some 75,000 years till about 235,000 years B.C. Acheulean man used his flints to scrape the skins of the animals he killed, so that he could wear them. The long story of man's use of his natural environment to provide himself with clothing had begun. It is a far cry from Acheulean man laboriously scraping the gristle from a deer-skin to clothe himself in an Ice Age winter, to the near extinction of the great crested grebe in Britain to provide hats for Victorian women, but the links of the chain are all there.

The High Terrace gravels have not affected the topography of London very much. In Inner London they are only found on the ridge of high ground between Camden Town and Highbury on which stand the Metropolitan Cattle Market and Holloway Gaol. The Middle Terrace, however, which was being laid down during the Acheulean period, is much more important. This is the gravel on which stand the greater part of the City and large areas of the East and West Ends of modern London. The City owes its very existence to the two low bluffs of Taplow gravel, separated by the Walbrook, which the Romans chose for their settlement. At about the same time were laid down the extensive deposits of "·brick-earth," which often overlie the river gravels. The brick-earth is rather like the wind-borne loess of the Continent and is fine-grained. It probably represents the fine dust blown by Arctic winds from the ice-sheet but deposited under water in the damper climate of England. It is found extensively in South Middlesex and South-west Essex, and being suitable both for the manufacture of bricks and tiles and for market-gardening has been much worked for both. Over considerable areas the brick-earth has now been entirely removed, especially after the Great Fire of 1666, when there was a great increase in the proportion of brick houses in London. Many of the fine Queen Anne houses in and around London owe their rich red bricks to the brick-earth of these deposits, blown from glaciers hundreds of thousands of years before.

Mammoths were common members of the London fauna at the time the Middle Terrace gravels and the brick-earth were being laid down, and many other of the mammals of the High Terrace period were also present. Hicks has recorded the excavation in Endsleigh Street, near Euston Station, of the remains of mammoth, deer, wild horse and a vole (*Arvicola*) in Middle Terrace gravel, and in other parts of London remains of the cave-bear (*Ursus horribilis*), woolly rhinoceros (*Rhinoceros tichorhinus*), hippopotamus (*Hippopotamus amphibius*), musk ox (*Ovibos moschatus*), and roe-deer (*Capreolus caprea*) have been found. In the brick-earth deposits at Ilford, Essex, many of these creatures and also the giant elk (*Cervus giganteus*), bison (*Bison priscus*), Saiga antelope (*Saiga tartarica*), now confined to the steppes of Russia and Western Asia, pouched marmot (*Spermophilus citellus*), tailless hare (*Lagomys spelæus*) and lemming (*Lemmus*) have been found.

The plants discovered at the Endsleigh Street site show that the flora of Acheulean England would have been more familiar than the fauna. Three kinds of buttercup or crowfoot were there (*Ranunculus aquatilis*, *R. sceleratus*, *R. repens*), the common chickweed (*Stellaria media*), tormentil (*Potentilla erecta*), mare's-tail (*Hippuris vulgaris*), water milfoil (*Myriophyllum spicatum*), knotgrass (*Polygonum aviculare*), spotted persicaria (*P. persicaria*), a dock (*Rumex obtusifolius*), three pond-weeds (*Potamogeton obtusifolius*, *P. crispus*, *Zannichellia palustris*), and a number of sedges including *Carex dioica* and *Eleocharis palustris*. As can be seen these are mainly the plants of a marshy habitat, and some of them such as the chickweed, polygonums, and dock, are plants which in later years succeeded in adapting themselves to cultivated habitats as "weeds." It is also noteworthy that there are no typically Arctic species, such as are often found in association with the Low Terrace gravels.

Some time during the Middle Terrace period of the Thames the Acheulean culture was itself replaced by the more advanced Mousterian culture, which began perhaps 235,000 years ago and lasted for about 150,000 years. Huge quantities of the flint implements of Mousterian man have been found on working floors in the Taplow gravels in Middlesex, notably at Acton and Stoke Newington ; at the latter place there is evidence of a late Acheulean settlement also. The men of the Mousterian culture of the Old Stone Age were of the type commonly known as Neanderthal, from the Rhenish valley where a skeleton dating from this period was found in 1857.

The weather continued cold throughout the Acheulean and Mousterian periods, becoming sub-arctic towards the end, so that the southern forms of animals and plants had to struggle to keep their places against the increasing differential advantage of the arctic forms. There was little, if any, change in the ecological status of man or in his importance as a biotic factor. Nor was there anything resembling a colonisation of London by Neanderthalers. The Mousterian settlement at Stoke Newington, which is within the present boundary of the County of London, was no different from those at Acton in Middlesex, Rickmansworth and Croxley Green in Hertfordshire, Dartford, Swanscombe and Greenhithe in Kent, or Grays in Essex. At Swanscombe, for instance, in a similar gravel terrace, Mousterian workshops have been discovered in association with abundant remains of mammoths and woolly rhinoceroses. These two great beasts must often have been hunted by the Neanderthalers of the lower Thames valley, and there is some evidence that men contributed to the eventual decline and extinction of both the mammoth and the wild horse, though climatic changes must have been much more potent. Man was still a relatively feeble instrument of ecological change.

Vulliamy has described Neanderthal man as having an unmistakably human form, which still retained certain ape-like features :

" His body was short and heavy. His head was large, with face and jaws more developed than the brain-box. The top of his head was low and flattened, his eyes were overhung by great ridges, his forehead was shallow and retreating, his neck enormously thick. His cheek-bones were flat, but his prominent muzzle and the complete absence of a chin gave him a brutish and sinister appearance. He walked with a shuffling gait, and was unable to assume a perfectly erect position."

This is clearly an advance on the Chellean man described above, his chin being his only really primitive feature. Like his predecessors, the Mousterian hunted and knew the use of fire, and was an even more skilled worker of flints. We know this from the great number of his implements that have been found in all parts of the lower Thames valley. He also had quite elaborate funeral ceremonies for burying his dead, providing them with food, weapons and shelter.

The third and latest terrace of the Thames gravels was laid down at a time that coincided with the last glacial irruption. The Flood Plain gravels are found in several broad strips in the valleys of the

PLATE 3

C. W. NEWBERRY

Docks and thistles on waste ground by flooded gravel pit at
Colne Valley, near Harefield

PLATE 4

ERIC HOSKING

Sand-martin colony in disused sand pit near Barnet By-pass, North Mimms

Thames and Lea. The Tower Hamlets, Limehouse and Bromley-by-Bow, stand on one such strip, and Fleet Street, the Strand and St. James's Park on another, separated from the Middle Terrace by a narrow tongue of London Clay lying between Piccadilly and St. James's. Westminster stands on an island of Low Terrace gravel that was formerly known as Thorney, while the whole flat plain from Chelsea to Chiswick is covered by these gravels, and in consequence was formerly one of the principal market garden areas of London. All over the London Area the gravels, both terrace and glacial, have been extensively dug for commercial purposes, and considerable areas of Middlesex, in particular, have been rendered useless for agriculture in this way. These flooded gravel-pits, as we shall see in Chapter 10., are now favourite resorts of aquatic birds and plants (Plate 2), and the ones which have appreciable cliffs have been turned to good use by sand-martins, which like to burrow their nesting-holes in soft sands and gravels. (Plate 4.)

We can tell from the flora and fauna of the Low Terrace deposits that have been excavated in London, and especially in the Lea valley, that the climate of those times was not unlike that of Lapland to-day. It is hard to imagine the Thames and Lea valleys as almost treeless wastes, moss-covered in the short Arctic summer, snow-covered in the long, cold winter. A stone's throw from Charing Cross, in Spring Gardens when the present Admiralty buildings were being put up, Abbott found the remains of a typical Arctic plant, the dwarf birch (*Betula nana*), its fossil leaves preserved in the Low Terrace gravels, together with a number of plants normally found in boggy or marshy conditions, such as *Ceratophyllum demersum*, pondweed (*Potamogeton*), and sedges (*Carex* and *Scirpus*). The extensive gravels of the Lea valley were thoroughly studied by Clement Reid, who has listed nearly seventy different kinds of plants from gravel-pits at Edmonton, Ponders End and Hackney Wick, the last-named within the London county boundary. Some are Arctic, such as the meadow rue (*Thalictrum alpinum*), thrift (*Armeria arctica*), a sorrel (*Oxyria digyna*), the dwarf birch and the Arctic willow (*Salix lapponum*). Others are familiar marsh plants to-day, such as the water-crowfoots (*Ranunculus aquatilis, R. hederaceus*), the lesser spearwort (*R. flammula*), the water-loving stitch-worts (*Stellaria aquatica, S. palustris*), blinks (*Montia fontana*), mare's-tail (*Hippuris vulgaris*), the small bur-weed (*Sparganium minimum*), and various pond-weeds and sedges. Yet others are common wayside and garden weeds of to-day, and must have been living then in quite different types of habitats from their normal present-day ones : the creeping buttercup (*Ranunculus repens*), bladder campion (*Silene cucubalus*), mouse-ear chickweed (*Cerastium arvense*), tormentil (*Potentilla erecta*), silverweed (*P. anserina*), marsh plume-thistle (*Cirsium palustre*), smooth hawksbeard (*Crepis capelloris*), orache (*Atriplex hastata*), knotgrass (*Polygonum aviculare*), curled dock (*Rumex crispus*), sorrel (*R. acetosa*), and stinging nettle (*Urtica dioica*).

Of the mammals that inhabited the Thames and Lea valleys when the Flood Plain gravels were being deposited, we also know something from the remains that have been dug up when the gravel was being worked by man thousands of years later. When the Admiralty buildings in Spring Gardens were being built, traces of mammoth, hippopotamus, rhinoceros, red deer and wild horses and bulls were found, together with the claw of a gull and remains of a tortoise (*Emys*), two beetles, a wood-louse and a good many molluscs. Abbott, who reported on this excavation, considered that after the gravels had been laid down the main channel of the Thames was diverted, and the gravels left as a bank, on which about two feet of sand were deposited. In this sand many molluscs, such as *Unio littoralis* and *Sphærium corneum* flourished. While this was going on the backwater was closed, and the gravel bank capped with sand dried, so that both plants and mammals, such as those mentioned above, were able to live on it. Later on the water rose again, and in the course of time a peat-bed was laid down, in which were found the remains of wild bulls and horses, sheep, wild boars, fallow-deer, hares and geese, as well as many molluscs.

The gravels of the Lea valley show a much more Arctic fauna, including mammoth, reindeer, bison, rhinoceros and lemming, and are considered by Warren to be typical of the Northern Tundra.

Such were the animal companions of man at the end of the Old Stone Age or Palæolithic period. To the beginning of the next period, the Middle Stone Age or Mesolithic, belongs the first truly human relic found in London, the skull of a woman found while the foundations of Lloyds' new building in Leadenhall Street in the City were being dug in 1925. This was the time of Cro-Magnon man, the first *Homo sapiens*, who gradually supplanted Neanderthal man all over Europe. He belonged to a different species—our own—and to a fine, tall race, averaging 5 ft. 9 in. in height, with long heads, large skulls, short, flat, broad faces, high cheek-bones and prominent chins. The Lloyds' woman belonged to the Aurignacian culture of the Mesolithic period, but of this and the other Middle Stone Age cultures in the London area we know practically nothing.

Nor do we know much more of the succeeding Neolithic period in the London area. The men of the New Stone Age arrived in this country somewhere between the eighth and sixth millennia B.C., when the climate was slightly warmer than it is to-day, pine-trees were replacing the formerly dominant birch over much of Southern England, and oak and hazel were also increasing. From the beginning of the sixth millennium B.C. the climate grew wetter again, and there was a substantial increase of oak, wych-elm and lime, and finally a great spread of alder, marking the fact that extensive areas of English woodland had become semi-waterlogged, in which state they were to remain until man cleared and drained them.

The men of the New Stone Age had not only partially domesticated certain animals, but cultivated the land, and so had to have definite settlements. There is no evidence, however, of any important neolithic settlements on the present site of London. If there were any, they must have been in the river valleys, as the clay lands in between would have been covered with an almost impenetrable waterlogged forest of oak and alder that could hardly be cleared with flint axes. It has been suggested that the immigrating neolithic people may have been important agents in the introduction and dispersal of plant seeds in the British Isles. Salisbury suggests that fool's parsley (*Aethusa cynapium*) and fumitory (*Fumaria officinalis*) were among the weeds now common in the London area that were once accidentally brought to this country by these invaders from the Continent.

It is to the warm, wet climate of the so-called Atlantic period from the sixth to the second millennium B.C. that we may attribute the establishment of the modern type of mixed oakwood, so typical of those areas of Middlesex and North Surrey on which the great mass of metropolitan London·was later to sprawl. According to Tansley, the vegetation now began to take on the main outlines of the character it would have to-day but for the interference of man. More or less the same may be said for the animal life. The reindeer had long gone with the retreating birches and pines. Mammoths and woolly rhinoceroses were hardly even distant folk-memories. Red deer, wild oxen, wolves and wild boars inhabited the great Forest of Middlesex, as they did still in Fitzstephen's day, some six thousand years later. The Thames must still have been broader and faster than it is to-day, with extensive mudflats at low tide, and innumerable small osier-covered eyots.

In the rather drier sub-Boreal period, which occupied most of the second and part of the first millennium B.C., we get for the first time an appreciable amount of beech pollen in the peat deposits, analysis of which has told us most of what we know of the changes in the climate and flora of Britain in the ten thousand years before the coming of the Romans. Thus the ancient controversy about the presence of the beech in England at the time of Julius Cæsar has been finally settled.

Neither in the Bronze Age, which began in Britain about 1800 B.C. with the arrival of the Beaker Folk, nor in the succeeding Iron Age which began about 600 B.C., was there, so far as existing evidence goes, any permanent settlement either on the present City area or on any part of the inner core of modern London. A few miles upstream, however, the remains of pile-dwellings from a Bronze Age village have been discovered at Brentford.

To the first millennium B.C. we can date a large increase in Britain of a tree that has had an important influence on the woodlands of the London area, which may, indeed, claim to be its metropolis in England. This is the hornbeam, which finds in the woods of Essex, Middlesex and Hertfordshire its north-westward limit in Europe. It is much valued for firewood, and at present is usually coppiced in the London area. In the nineteenth century it was the determination of a Loughton villager to maintain his ancient right to lop the pollard hornbeams in Epping Forest (Plate 5) that was indirectly responsible for the Forest being saved as a public open space. (See p. 222.) It is of interest to note that the Hampton Court Maze is made of high hornbeam hedges.

Up to the middle of the first century B.C., when Cæsar made his first expedition to Britain, the influence of man on the animals and plants of the London area fell mainly under the headings of domestication and destruction for safety, food, clothing, fuel and cultivation. Though horses, cattle, sheep, swine, fowls and dogs all formed part of the animal community centred on man by this time, it must not be supposed that the actual wild individuals found in the lower Thames valley had been domesticated. It is much more likely that the original

PLATE 5

ERIC HOSKING

Pollard hornbeams in Epping Forest, near Loughton

ERIC HOSKING

The river wall at the confluence of the Ingrebourne with
the Thames at Rainham, Essex

ERIC HOSKING

The Upper Pool at sunset: London Bridge in the background

stock had been imported ; this applies also the chief members of the domesticated plant community, notably the grains.

For his own safety and that of his stock we may suppose that from his very first emergence as a separate species, man has killed predatory birds and beasts. Palæolithic man must often have slain cave lions, cave bears and wolves in the sheer battle for survival, and his neolithic successors had the added motive of herds and crops to protect. Many animals were also hunted for food, notably deer and fish, and when mammoths were available a dead mammoth must have provided meat for a tribe for many days. Wild fruits, berries and herbs were gathered for food, but not on a scale to affect seriously the ecological balance. Throughout the Stone Age animals were killed for the clothing their skins afforded and the fuel oil obtainable from their fat. Often the same animal, say a bear, would be killed because it was a danger to man and his domestic animals, because it was good to eat, because its skin could be used for clothing, and because its tallow could be used in a primitive lamp. For fuel for fires, trees and brush-wood must have been destroyed also throughout the prehistoric period, and towards the end of it substantial areas of the lower lying ground near the river had probably already been cleared by human agency. The need for wood for building huts would have been a rather minor factor contributing to the felling of woodland. More land would have been cleared for cultivation, and for the actual sites of villages.

So far as the area now occupied by London is concerned, it is fairly safe to say that all human activities from the remote Chellean period to the arrival of the Romans had little impact on the natural animal and plant communities of the lower Thames valley beyond the clearing and primitive cultivation of a few acres of land and the erection of a few clay and wattle huts ; though as suggested above a fairly large area of woodland near the river may have been cleared for other purposes. We may confidently say that the united efforts of the men of the million or so years of the Stone, Bronze and Iron Ages combined had less effect on the natural communities than their twentieth century successors had in a single week of " development " ; less even that the introduction some centuries later of that exceedingly destructive rodent, the rabbit.

References for Chapter 2

Abbott (1892), Dewey (1932), Fitter (1941*b*.), Godwin (1940, 1941), Hall (1934), Hicks (1892), Leach (1909), Reid (1915), Rudler (1913), Salisbury (1927), Tansley (1939), Vulliamy (1930), Warren (1912, 1915), Willatts (1937), Winbolt (1943), Woodward (1901), Wright (1937).

ROMANO-BRITISH LONDON

WHETHER or not London existed before the Romans, we simply do not know. Though London's name is undoubtedly of Celtic origin, there is no trace of a pre-Roman settlement of any consequence within the area which for nineteen centuries since the arrival of the Romans had been the core of London, the two low gravel-capped hills on either side of the Walbrook. Moreover, the whole of such evidence as exists does not carry the possible date of the foundation of London back for more than a decade or so before the Claudian invasion of A.D. 43. It is true that in the hundred and fifty years before this the Belgæ had crossed the Channel and founded towns on or near the sites of the modern Colchester, St. Albans and Winchester, but it is very hard to believe that they could have made any settlement of comparable importance on the site of London without any evidence, either archæological or literary, coming down to posterity. There can be no doubt that the Royal Commission on Historical Monuments was right in its conclusion, that :

> " On all grounds it must be admitted that, whilst the possibility of some pre-Claudian occupation of the site of London cannot yet be finally dismissed, there is at present no valid reason for supposing that London existed prior to A.D. 43."

What is certainly known is that very soon after the Claudian invasion the Romans founded a town on the site of the present city, which rapidly became a thriving port. Only seventeen years after the arrival of Aulus Plautius, whose camp in the autumn of 43 may even have begun London, the great British warrior-queen, Boudicca, swept down on the Roman city and sacked it, evoking from Tacitus the first mention of London in written history :

"At Suetonius mira constantia medios inter hostis Londinium perrexit, cognomento quidem coloniæ non insigne, sed copia negotiatorum et commeatuum maxime celebre."[1]

Even at this early date London must have had a human population of at least ten or twenty thousand, for Tacitus a few lines later mentions that the combined population of Colchester, London and Verulam (St. Albans) was 70,000.

We have some idea of what the immediate environs of London must have looked like when Roman spades and mattocks were turning the Thames-side sods to lay the foundations of Londinium (Map 2), but we may perhaps start with the statement of what they were not like. As a result possibly of a misinterpretation of Cassius Dio's reference to the Britons retiring before Aulus Plautius to the Thames " at a point near where it empties into the ocean and at flood-tide forms a lake," there is a widespread impression that Roman London was a kind of island in a marsh or lagoon that stretched over wide areas of modern Central London, wherever alluvium is now shown on the geological map. Recent research, however, suggests that this idea, assiduously preached by Sir Walter Besant and others, is wide of the mark. We now know that the level of London has fallen something like fifteen feet in the millennium and a half since it was Londinium. The layer of Roman deposits, 1 to 10 feet thick, is now anything from 9 to 22 feet below the level of the ground ; at Tilbury, for instance, remains of Romano-British huts can still be seen on the foreshore some 13 feet below high-water mark. This means that all the formerly marshy ground of South London, some of which is actually below sea-level to-day, was in Roman times well above the reach of any but the highest tides. In fact, it is doubtful if the Thames was normally tidal above the present site of London Bridge in the first century B.C., so that Cæsar, wherever he forded the Thames, most likely did so above the then tidal limit, and the lagoon referred to by Cassius Dio must have been some miles downstream.

It has sometimes been suggested that it was the Romans who first embanked the Thames at London (Plate 6a) but the Royal Commission quotes with approval Spurrell's judgment that " of banks against the tide in the district below Purfleet there are none surviving of the

[1] "Now Suetonius with remarkable determination marched straight through the midst of the enemy to London, a place not indeed distinguished by the title of colony, but crowded with traders and a great centre of commerce."

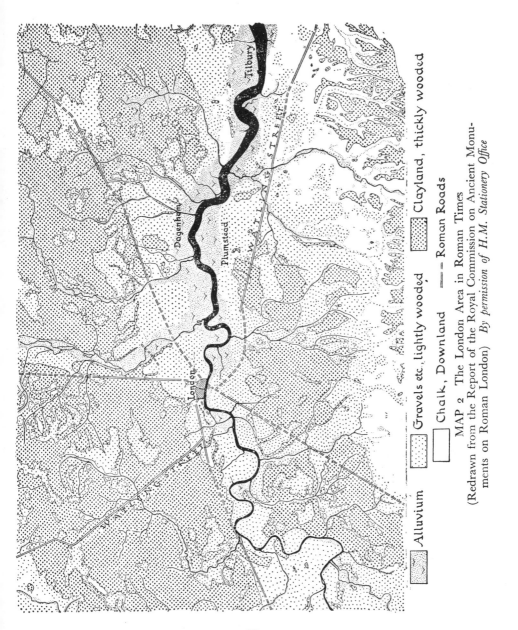

Alluvium

Gravels etc., lightly wooded

Clayland, thickly wooded

Chalk, Downland

= Roman Roads

MAP 2 The London Area in Roman Times
(Redrawn from the Report of the Royal Commission on Ancient Monuments on Roman London) *By permission of H.M. Stationery Office*

29

Roman period, while above that place none or but the slightest ones
were needed, and no signs of any can be found." Additional evidence
of the lack of tidal flow in the Thames below London two thousand
years ago is found in the presence of the remains of a forest of birch,
elm, oak, hazel and yew in the peat of the marshes between Greenhithe and Purfleet, and nearer London. Though the yew is intolerant
of water and will not live in saline conditions, a forest of yew-trees
existed at a time not very remote from the arrival of the Romans on
the present site of the marshes on either side of the river below Dagenham and Plumstead, showing that the level of high tide was then
probably some 15 feet below that of the present land-surface.

Though Londinium was not surrounded by the miasmatic lagoon
so graphically described by Besant, the river was almost certainly
broader and shallower than in modern times, even before the building
of the Embankments. The gravel terraces flanking the Thames on
either side probably supported a light scrub, with scattered trees, and
alder and willow along the banks of the Walbrook, Fleet, and other
streams. On the clay lands to the north and north-west the site of
London must have been surrounded by a dense and probably waterlogged oak-forest, the Forest of Middlesex as its remnants were later
called ; in its dense tangled undergrowth of hazel, thorn and bramble
wolves and other fierce beasts lived. The two low hills of gravel on
either side of the little Walbrook, whose course can still be traced down
the narrow City street that bears its name, were conveniently situated
at the lowest point where the river could be bridged. It was therefore
natural that a settlement should grow up there when the Romans did
build the first London Bridge.

When the town was built, and rebuilt after the first Fire of London
in 60, it must have had the same destructive effect on the plant communities of the immediate neighbourhood that occurs whenever an
urban settlement is made. The plants growing on the sites of the roads
and buildings are completely destroyed, while those on the sites of
gardens have to compete with man's introduced plants and his efforts
to eradicate the native flora as " weeds." The Romans also brought
their domestic animals with them, horses and dogs, cattle, sheep and
goats, some of which they would have been able to acquire from the
Britons. For South-east Britain in the first century A.D. was a relatively
prosperous agricultural area, with rich corn-lands in Essex, Kent and
Hertfordshire. A hundred years previously Cæsar had already noted

a large human population (" hominum est infinita multitudo "), with many farmsteads like those in Gaul and numerous cattle (" pecoris magnus numerus "). There was thus a rich food-growing hinterland to support Londinium, which eventually reached a population of some 50,000 and must have drawn its food-supplies from considerable distances. Insofar as the demands of the Romano-British Londoners for meat and corn and dairy-produce caused more land to be turned over to cultivation, London began to have its first influence on the surrounding country. From now on London's impact on its neighbouring fields was to be a constant progression of cultivation followed by building.

Refuse heaps in London have yielded the bones of oxen, pigs, sheep, horses, goats and dogs among domestic ; red and roe deer, hares and birds among wild animals. Possibly in London as in Silchester, the raven, a favoured scavenger, may have lived in a semi-domestic state ; in the latter place the bones of ravens were second in abundance only to the domestic fowl. Fragments such as these are all the information we have on which to build a picture of the inter-relations of the human community and the animals and plants of Roman London.

There must have been a fairly large area of open space within the walls that were erected round the 330 acres of the city in the fourth century, at a time when Londinium temporarily assumed the high-sounding name of Augusta. It is pretty certain that apart from some villas along the river's edge near the present site of St. Clement Danes, all the built-up parts of Londinium were within these walls. Outside the walls there must have been a certain amount of cultivated ground and some clearance of the woodland on either side of the main roads that linked Londinium to the rest of Britain, particularly Watling Street, which ran through the dense oak forest to Verulamium (St. Albans) and on to Deva (Chester).

The small stream later known as the Walbrook ran through the heart of Londinium. The flanks of its valley were built over at an early stage in the development of the city, and the uninhabited area represented by its valley floor must have been quite small. Low walls were built to contain it, and the valley gradually filled up with rubbish. The construction of the city wall blocked up the northern inlets of the stream, and these had to be culverted by the Romans. When the culverts finally got blocked, mainly after the Roman period, the great marsh of Moorfields was formed, and persisted throughout the Middle Ages.

One introduction, if a temporary one, that the Romans made is of some interest. The elephants that Aulus Plautius brought with him in 43 were the first to set foot on British soil for a good many thousand years, and indeed were the absolute first of their species, while it would be many centuries after the end of the Roman occupation before any more were seen in the streets of London. Various other introductions, accidental and otherwise, by the Romans have resulted in more permanent additions to our fauna and flora ; for instance, Salisbury suggests that the common cornfield weed penny-cress (*Thlaspi arvense*) was probably introduced at the time.

There is as much controversy and lack of knowledge about the end of Londinium as about its possible Celtic beginnings. Was it sacked by the invading Saxons and laid waste ? Was it just abandoned by the retreating Britons and not occupied by the invaders ? Or did it continue to exist as an urban community right through the two centuries, from 429 to 604, when we have no written record of it ? The Royal Commission lends its support to the view that some form of continuity was maintained :

" There is no substantial reason why the few surviving walled cities should not have remained as islands in the flood. . . . The walled towns of the south-east, out of reach of the Picts and out of mind of the Saxons, may thus be thought to have lingered on almost as ' reservations ' for the secondary Romano-British population. The silence of history in regard to them is probably just ; London in the year 500 can have mattered little to anyone save a few decivilised sub-Roman Londoners."

It would at least be in keeping with the later history and traditions of London if the Londoners of the fifth century had come to terms with the Saxon invaders, as they later did with William the Conqueror, and retained their semi-independence as a useful trading post amid a welter of little tribal kingdoms. It is curious that the city has never been a part of the county of Middlesex, which surrounds it except for its river frontage, and that right into the Middle Ages the Londoners retained special hunting rights in the Forest of Middlesex and in the Chilterns, which it would have been rather odd for an ordinary Saxon

PLATE 7

W. SUSCHITZKY

Feeding the pigeons in front of St. Paul's Cathedral

ERIC HOSKING

A rabbit warren in Hainault Forest, Essex

ERIC HOSKING

A badger holt at Loughton Camp, Epping Forest

township to have acquired (cf. p. 52). On the whole, it seems more probable that some people continued to live in London right through the Dark Ages than that at any time the city was wholly given up to be " a possession for the bittern."

References for Chapter 3

Besant (1892), Cæsar (1st cent. B.C.), Cassius Dio (1st-2nd cent. A.D.), Gomme (1914), Home (1926), Loftie (1883), Major (1920), Ritchie (1920), Royal Commission on Historical Monuments (England) (1928), Salisbury (1927), Spurrell (1885), Tacitus (1st cent. A.D.), Vulliamy (1930).

CHAPTER 4

MEDIEVAL LONDON

". . . London, small and white and clean,
The clear Thames bordered by its gardens green."
WILLIAM MORRIS, *The Wanderers*.

WHEN London at length emerges from the fog of tradition into the light of written history in A.D. 604, it is as the seat of the East Saxon bishopric. For the next two and a half centuries we may picture a small Saxon community living within the vast crumbling shell of Roman London, often in the patched-up remains of Roman houses. Much of the 330 acres enclosed by the Roman wall must have been at this time, and for many centuries, open space. As late as the reign of Henry II we find it recorded that half London was waste. Not for many years was there any question of London breaking its bounds and spreading over the Middlesex countryside beyond the wall. But when it once began there was no stopping it.

In the middle of the ninth century there fell on the city one of those cataclysms that were to be a regular feature of London life till the Great Fire of 1666 wiped out almost all that remained of medieval London. In 851 London was sacked by the Danes, to such effect that twenty-one years later the city had hardly begun to be rebuilt, for it is recorded that in 872 the Danes came again and encamped in the ruins. When the city was rebuilt, through the energy of that greatest of English kings, Alfred, it was no longer a brick-built and therefore partially fireproof London, but a wooden-walled, thatch-roofed, highly combustible medieval London. Thereafter the city was laid waste by fire with almost monotonous regularity, in 982, 1077, 1087 and 1136, for example, while in many other years substantial damage was caused by fire to its great cathedral of St. Paul and other buildings. In 1203 it was said that birds were seen flying in the air, with burning coals in their bills, which set light to many houses in London, but it is hard to believe that this represents anything but the bright idea of some thirteenth century journalist in the silly season. Gurney suggested it might refer to the mischievous habits of the jackdaw, but there does not seem to be any other record of jackdaws or other birds or even beasts, carrying burning brands ; there were Samson's foxes, but they were hardly willing agents.

The constant laying waste of large areas of the city must have made the aspect of medieval London often a good deal more like the blitzed London of 1945 than most people realise. Added to the fires were great gales, like that of 1091 which came from the south-east, blowing down 600 houses and unroofing several churches.

Only towards the end of the Middle Ages did London begin to expand beyond its walls, the first mention of the city's jurisdiction extending to the suburbs under its walls occurring in a charter of Henry III in 1268. Early in the following century the population began to increase to such an extent that the land occupied by the present wards of Aldersgate Without, Bishopsgate Without, Cripplegate Without and Farringdon Without was incorporated with the City. This is the only time that the City of London has annexed any part of the County of Middlesex, that part of the City which lies within the walls never having at any time been part of Middlesex.

For a description of medieval London every historian of the city is indebted to William Fitzstephen, the Cockney-born monk of Canterbury, who prefixed to his life of St. Thomas Becket, written in the twelfth century, an account of his own and Becket's native city, which, he said, " amongst the noble and celebrated cities of the world . . . is one of the most renowned, possessing above all others abundant wealth, extensive commerce, great grandeur and magnificence." Fitzstephen's description of the environs of the city is worth quoting in full :

" On the east stands the Palatine tower [i.e., the Tower of London], a fortress of great size and strength, the court and walls of which are erected upon a very deep foundation, the mortar used in the building being tempered with the blood of beasts. On the west are two castles strongly fortified ; the wall of the city is high and thick, with seven double gates [i.e., Ludgate, Newgate, Aldersgate, Bishopsgate, Aldgate, Tower postern, Bridgegate], having on the north side towers placed at proper intervals.

London formerly had walls and towers in like manner on the south, but that most excellent river the Thames, which abounds with fish, and in which the tide ebbs and flows, runs on that side, and has in a long space of time washed down, undermined, and subverted the walls in that part, On the west also, higher up on the bank of the river, the royal palace rears its head, an incomparable structure, furnished with a breastwork and bastions, situated in a populous suburb [i.e., Westminster], at a distance of two miles from the city.

" Adjoining to the houses on all sides lie the gardens of those citizens that dwell in the suburbs, which are well furnished with trees, spacious and beautiful.

" On the north side too are fields for pasture, and a delightful plain of meadow-land, interspersed with flowing streams, on which stand mills, whose clack is very pleasing to the ear. Close by lies an immense forest, in which are densely wooded thickets, the coverts of game, stags, fallow-deer, boars and wild bulls. The tillage lands of the city are not barren gravelly soils, but like the fertile plains of Asia, which produce abundant crops, and fill the barns of their cultivators with ' Ceres's plenteous sheaf.'

" There are also round London, on the northern side, in the suburbs, excellent springs ; the water of which is sweet, clear, and salubrious, ' 'mid glistening pebbles gliding playfully,' amongst which Holywell, Clerkenwell and St. Clement's well, are of most note, and most frequently visited, as well by the scholars from the schools, as by the youth of the city when they go out to take the air in the summer evenings. The city is delightful indeed, when it has a good governor."

THE EFFECTS OF BUILDING

Up to the end of the Middle Ages the built-up area of London, apart from the houses of the more substantial citizens surrounded by large gardens, did not extend beyond the present boundaries of the City, though there were two outlying settlements, Southwark, the bridgehead at the south end of London Bridge, and Westminster, ·clustered around the royal palace and the great monastery on the patch of gravel known as Thorney Island. Several other great monas-

PLATE 9

ERIC HOSKING

The New River and the North-East Reservoir at Stoke Newington

Aldenham or Elstree Reservoir, Herts.

Walthamstow, No. 5 Reservoir, showing the heronry island

teries and religious houses lay in the fields without the walls of London, notably the Abbey of Bermondsey, where the Benedictines of the reformed Cluniac rule were settled in the marshes by Aylwin the Child in 1082, the Charterhouse by Smithfield, where in 1371 Sir Walter de Manny established the Carthusians, and the Priory of the Knights of St. John of Jerusalem, founded at Clerkenwell about 1100. Otherwise the whole area of the present boroughs of Holborn, Finsbury, Shoreditch, Stepney, Bermondsey and Southwark and of the City of Westminster, which abut on the City of London and in 1931 had a combined population of nearly 800,000, was open country, largely under cultivation.

The little river Walbrook flowed openly through the city in the Middle Ages, falling into the Thames at the small dock of Dowgate, which still exists. The city's other river, the Fleet or Holebourne, was of quite respectable dimensions in medieval times, and was navigable by ships of small draft as far up as Holborn Bridge, near the present site of Holborn Viaduct. Later the Fleet became notorious as an extremely noisome open sewer.

We know from Fitzstephen that the part of the Roman wall facing the river had by the twelfth century been destroyed by erosion and neglect. This wall stood a good distance back from the present waterfront, the area between which and Upper Thames Street was reclaimed from a tidal marsh mainly in the Middle Ages. The rest of the Roman wall survived the sackings of the Danes and the numerous fires, only to succumb over the greater part of its length to the commercial rapacity of the eighteenth century.

As we have seen, even the area within the city wall was not wholly built over in the Middle Ages. As late as the reign of Henry II in the latter part of the twelfth century a large area was still lying waste from previous devastations, and throughout this period there were orchards and gardens right in the heart of the city. In the earlier part of the Middle Ages there was possibly pasture and even arable land within the walls. In the ninth and tenth centuries, when commercial London still clustered closely about the bridge and Eastcheap market, a large area of open land west of the Walbrook was in the hands of the King and the Cathedral authorities. This land the royal and ecclesiastical landlords proceeded to develop by granting sokes, feudal equivalents of leases, just as if they had been granting tracts of fen and forest land to one of the great monasteries for development. The area enclosed by London Wall, Broad Street, Lothbury and Coleman Street was probably a marsh as late as the middle of the twelfth century.

The streets were unpaved and mostly uncleansed till the early part of the fourteenth century, hence the large numbers of carrion birds, notably kites and ravens, that frequented the city. Houses during the greater part of the Middle Ages were made of wood, with thatched roofs. Repeated laws tried to secure the partial use of stone, with slate or tiled roofs, to prevent the recurring fires, but at the beginning of the fourteenth century some houses were still thatched. It seems unlikely that the relatively small amount of timber and thatch required to construct the houses of medieval London can have had any marked effect on the woods or reed-beds nearby. Insofar as woods were cut down or reed-beds destroyed to provide these materials, it is probable that the advance of cultivation would in any case have achieved the same result not long after.

THE EFFECTS OF CULTIVATION AND DOMESTICATION

Next to the use of land for building and streets, the most massive influence of civilised man on his natural environment has always been the cultivation of large areas of land to grow food for himself and his domestic animals, coupled with the actual domestication of both animals and plants, mainly for the purpose of having food readily at hand. In the London area the direct effects of domestication have probably at all periods been small. Nearly all the domestic animals in London in historic times have been introduced from stocks domesticated elsewhere. The single possible exception is the swan. As regards plants, though some of the wild stocks growing in and around London must have been used for domestication, we have no specific evidence that, for instance, the strawberry plants which grew the fruit referred to in the fifteenth century poem, *London Lyckpeny* :

> " Then unto London I did me hye,
>   Of all the land it beareth the pryse
>   Hot pescodes one began to crye
>   Strabery rype and cherryes in the ryse. . . ."

were in fact domesticated from the wild strawberries (*Fragaria vesca*) whose descendants can still be found on hedge-banks in Middlesex.

It is unfortunate that Domesday Book, which has left us such an admirable picture of eleventh century agriculture in the rest of the County of London, contains no record for the City area, though we know that large parts of it were open enough to be farmed at that time. In the manor of Westminster, which covered about half the area of the modern City of Westminster, there were eleven ploughs, which meant about 1300 acres of arable land, since each team of eight oxen was reckoned able to cope with about 120 acres in a year. Westminster also contained pasture for the cattle of the vill, which would include the eighty-eight plough oxen, wood for a hundred pigs, and four arpents (rather more than one acre) of newly planted vineyard. All this was farmed by nineteen villeins and forty-two cottagers, who had gardens.

In the Bishop of London's great manor of Stepney on the eastern side of the City there was land for twenty-five ploughs. This manor covered roughly the area of the present boroughs of Stepney, Poplar and Bethnal Green, so that some 3000 of the present acreage of 4800 was then under arable cultivation. There was also meadow for the plough teams and pasture for the cattle of the vill, woodland to feed 500 pigs, and scrub woodland providing wood for hedges and fences, as well as four mills. The manor of Stepney was farmed by sixty villeins and forty-six cottagers.

Westminster and Stepney were fairly typical of the farming of the rest of the modern County of London at the end of the eleventh century. Kensington had three arpents of vineyard and wood for 200 pigs. Totenhall, from which Tottenham Court Road derives its name, had woodland for 150 pigs. The manor of Eia (Ebury) on the site of Hyde Park produced hay worth sixty shillings a year. Holborn had a vineyard.

The immediate neighbourhood of London was evidently fairly intensively cultivated at the time of the compilation of the Domesday Survey in 1086. It must already have long lost most of the tangled thickets of alder and osier and the dense oak forests through which Cæsar had to force his way after fording the Thames over a thousand years before. Much woodland must have been cleared and put under the plough by Romans, Saxons and Vikings before William the Norman came to make his tax-gathering survey. Much marshland bordering the river must have been drained and protected from the tides by the low sea-wall that still stretches right down the Thames estuary, to

Foulness on the Essex shore, and to the Medway on the Kentish shore.
(Plate 6a.)

The land reclaimed and tilled so laboriously through many centuries was cultivated on one of two main systems. In the two-field
system half the land lay fallow while the other half was sown either
with autumn wheat or rye or with spring corn, oats or vetch. In the
three-field system only one-third of the land lay fallow, while the two
other thirds were sown one with autumn corn and the other with
spring corn. The practice of fallowing must have led to an enormous
proliferation of weeds, but it was the only way then known of ensuring
that all the fertility of the soil was not exhausted. After the harvest
the livestock of the manor was turned out on to the stubble to manure
it and help to restore its fertility.

The chief crops grown in England in the Middle Ages were wheat,
barley, rye, oats, vetches, beans and maslin, a mixture of wheat and
rye much favoured in the south. In a good year a threefold return
could be expected, or $7\frac{1}{2}$ bushels per acre. The chaff and bean-bines
were used as cattle-food, and together with some hay were all that
was available to keep cattle alive through the winter. The livestock
of the average lord of the manor included draught oxen, cows for
milk and meat, sheep, pigs and poultry. Ordinary villagers naturally
had less livestock than the lord, but most had a cow and a few pigs
and poultry. The lord could probably spare some grain, skim milk
and brewing residues for his pigs, but mostly, like everybody else's, they
had to forage for themselves in the woods, picking up acorns and
beech-mast in the autumn. Most manors had some woodland for
their pigs, and the whole county of Middlesex contained enough
woodland to support some 20,000 swine, the manors of Enfield
and Harrow each having pannage for 2,000. In the City pigs ran
semi-wild about the streets through most of the Middle Ages,
performing a useful function as scavengers. They were not, of
course, the rotund baconers and porkers that grace Smithfield
Cattle Show to-day, but sharp-faced, razor-backed creatures hardly
distinguishable from their wild brethren that still inhabited the great
Forest of Middlesex.

The poultry mostly fended for themselves, and must have been just
as different from our portly modern wyandottes and Rhode Island
Reds as were the swine from Large Whites and Berkshires. Ducks and
geese were also kept, and it is possible that some of these were domes-

Plate I

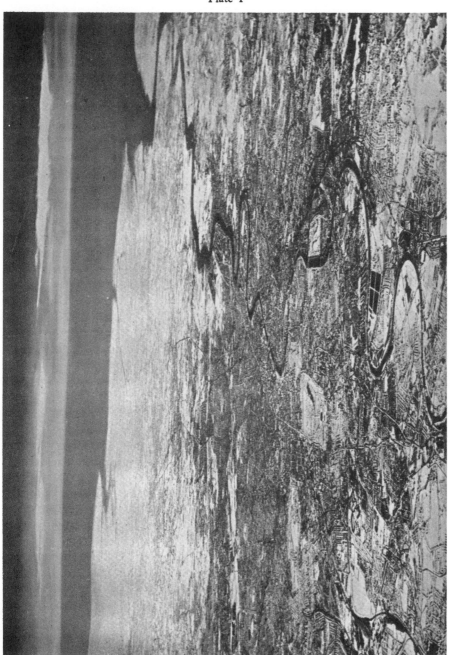

Aerial view of London and the Thames Estuary

Plate II

Central London from a tall building near St. Paul's

ticated from the mallards and grey-lag geese (*Anser anser*) of the Thames marshes.

Direct domestication is most likely to have happened with the Thames swans. It is now known that the legend of the introduction of the mute swan into England from Cyprus in the reign of Richard I, if it had any basis in fact, was a case of bringing coals to Newcastle, for swans already existed in a semi-wild condition in the Lincolnshire Fens long before the lion-hearted king visited Cyprus. We know that the grey-lag geese of the Lincolnshire and East Anglian Fens were gradually domesticated, and it seems reasonable to suppose that the same process was applied to the swans of the Fens and other parts of England. Dr. Norman Ticehurst has traced the custom of swan-keeping back to before 1186, and by the end of the Middle Ages swans were exceedingly numerous on the Thames at London. The secretary to the Venetian Ambassador wrote in 1496-7, " it is truly a beautiful thing to behold one or two thousand tame swans upon the river Thames, as I, and also your Magnificence have seen," and fifty years later the secretary to a Spanish duke remarked that he had never seen a river so thickly covered with swans as the Thames.

The earliest record of swan-keeping on the Thames relates to Buckinghamshire in 1230, but sixteen years later Henry III issued a mandate to the Sergeant of Kennington which referred to swans belonging to the King and to the Knights Hospitallers of Hampton, Middlesex. This mandate contains the first reference to the custom of dividing a brood of cygnets equally between the owners of the parent birds instead of the owner of the dam keeping them all, as happens with all other domestic animals. In 1249 occurs the first reference to the use of swans for food in London, and a few years later we get evidence that swans were not only well established items of diet, but were normally bought and sold in the open market. In the reign of Edward III the price of a swan was four or five shillings, nearly ten times that of a goose or mallard, and three or four times that of a pheasant.[1]

A large body of law and custom grew up around the London swans, and Ticehurst has described 135 different marks with which private owners used to mark their birds, while it is known that as many as 630 marks were used between 1450 and 1600. The annual swan-

---

[1] In the reign of George VI swans were again, under the special conditions of war-time, to be seen in London poulterers' shops, at prices of 40s.-55s. each.

upping, when the young birds are caught and their bills marked, still takes place on the Thames, even in wartime. Since the end of the eighteenth century, however, swan-rights have been exercised on the Thames only by the Crown and the Dyers' and Vintners' Companies. In the Middle Ages several Acts of Parliament restricted the ownership of the Thames swans. In 1483, for instance it was laid down that only persons with freehold estate of at least five marks a year could own them, but it is clear from the records that large numbers of swans were kept by persons not entitled to do so, and the crime of swan-stealing was rife. In 1357 Edward III was obliged to concede to private persons a grant of all unmarked swans on the river between London and Oxford for seven years.

Orchards and gardens were cultivated as well as arable land. Apples, plums and cherries grew in the orchards, and onions, leeks, garlic and cabbages in the vegetable gardens, while the more specialised herb gardens grew a variety of herbs, such as sage, rosemary, marjoram and thyme. It is remarkable how few of our modern vegetables seem to have been at all widely grown and eaten in the Middle Ages. Though Mayster Ion Gardener, writing in 1440, mentions seventy-eight different kinds of herbs and vegetables, most of them were only grown in a few specialised monastic and other great gardens. The average Londoner was like Chaucer's Sompnour,

" well loved he garlike, onions and lekes."

The large number of herbs grown, many of them, such as chickweed (*Stellaria media*), borage (*Borago officinalis*), and fennel (*Fœniculum vulgare*), having since reverted to the status of weeds, was symptomatic of the poor quality of the meat of the time. Those herbs which were not used to make high or salt meat palatable were used for simples. Many herbs, of course, were gathered wild.

One item in the Domesday Book, that of the vineyards, is particularly interesting, in view of the mistaken but widely held view that grapes cannot ripen out of doors in the English climate. We do not know, of course, what was the quality of the wine produced—who knows but that Holborn 1086 may not have been a famous vintage in Norman England?—but except in the very worst summers a good crop would ripen. What eventually killed the English vineyards was not so much a deterioration of the climate as the facilities for importing better wines from abroad, coupled with the dissolution of the monasteries, which had been the chief centres of viticulture in England, so that their accumulated skill was lost. Alas, no record has come down to us of the flora and fauna of an English vineyard[1].

MEASURES FOR THE SAFETY OF MAN AND HIS STOCK

In the Middle Ages the steps taken to protect domestic animals and even man himself from personal danger were still an appreciable biotic influence. The most important general effect was that occasioned by cutting down the forests which harboured both wild beasts and a variety of highwaymen and robbers. We have seen that the Romans cut down the forests on either side of their main roads, and we may suppose that after they left the woodland recolonised the margins, again to provide hiding places for wolves, human and lupine. In early Saxon days the woods of Middlesex were infested by robbers, and honest men passed through them with loud shouting and blowing of horns. Matthew Paris tells us that Leofstan, Abbot of St. Albans in the reign of Edward the Confessor, cut through the thick woods that stretched from the edge of the Chilterns nearly up to London, smoothed the rough places, built bridges and levelled the rugged roads, which he made safer, " for at that time there abounded throughout the whole of Ciltria [i.e., the Chilterns] spacious woods, thick and large, the habitation of numerous and various beasts, wolves, boars, forest bulls [' *tauri sylvestres* '] and stags." These wild bulls are evidently the same as those mentioned by Fitzstephen over a century later, after which we hear no more of them. It is quite likely that they were white, like the wild cattle that still survive in parks at Chillingham, Northumberland, and until 1905 at Chartley, Staffordshire.

By the tenth century some of the more fearsome beasts of prey, notably the lynx and the bear, had been eliminated in the London area. Wolves, however, still survived, and were not exterminated till some centuries later, though we do not know exactly when the last wolf was killed near London. The extermination of the wolf in England was entirely due to its destructiveness to man's herds and flocks, and in hard winters to man himself. Even in the 1930's little snippets of

[1] The Great Vine at Hampton Court (Plate III) was planted in 1768. An offshoot of this vine, planted in the Royal Gardens at Windsor in 1775, has long outgrown its progenitor.

news appeared almost every year in odd corners of the British press
about the wolves having come down into the Hungarian plains. Seven
or eight hundred years ago such items of news must have been common
in the ale-houses of London. Wolves must often have descended from
the high Chilterns and North Downs, or even from the wooded heights
of Hampstead and Dulwich, to devastate the flocks and herds grazing
within sight of the City walls, or carry off children left to play unguarded
out of earshot of their parents.

Within the immediate vicinity of the City many other predatory
beasts and birds, foxes, polecats, martens, wild cats, hawks, buzzards,
perhaps even eagles, would have to be kept down to protect the sheep-
folds and the poultry yards that fed the citizens. Later on these were
all to be destroyed in the name of sport, but in the Middle Ages they
were still sufficiently abundant to be a menace to domestic animals.

The Utilisation of Wild Animals and Plants for Food

In the Middle Ages the taking of wild animals and plants for food
for man and his domestic animals was still an important activity in
the London area. Indeed, it is difficult for modern Londoners, who
live to such a large extent on the produce of soil on the other side of
the world, to realise how largely their poorer ancestors were dependent
on going out and foraging for themselves in the woods and fields near
the City's gates.

The most important method of bringing wild animals into the
cooking pot was hunting, but as this was already beginning to turn
from a purely food-getting expedition into a popular as well as a
royal sport, we may defer consideration of it for the moment. It is
not out of place, however, to refer to the introduction of the rabbit
(Plate 8a), an event which possibly had a more profound effect
on the vegetation of the British Isles than any other single human
act.

It is now generally held that the rabbit (*Oryctolagus cuniculus*) died
out in the British Isles during the Ice Age, so that its later appearance
was a re-introduction. It is not easy to assess the value of discoveries
of the remains of rabbits—their bones have been found in neolithic
kitchen middens in Scotland—owing to their propensity for burrowing
and consequent ability to die several strata lower than their contem-

poraries. However, it seems fairly certain that there were no rabbits in either Roman or Saxon Britain. Cæsar mentions hares, fowls and geese in Britain, and might be expected to have referred to rabbits also if they had been present—though if he missed so prominent a tree as the beech, why not also the relatively insignificant rabbit? A more substantial reason for assuming the absence of rabbits in Britain between the first century B.C. and the fifth century A.D. is the absence of any of their remains from excavations of kitchen refuse in Romano-British settlements. There are, in fact, no pre-Norman traces of or references to the rabbit in Britain, while after the Conquest " conies " are often mentioned in the records.

In default of any direct evidence to the contrary, therefore, we can assume that the Normans introduced the rabbit into England. There is no doubt that it was then, and for many centuries later, regarded as a purely utilitarian animal, valued for its flesh and skin, but wholly unworthy to be a beast of the chase. It was at first treated much as musk-rats and silver foxes are to-day, being carefully preserved in warrens, and on islands, such as Lundy and Skokholm at the mouth of the Bristol Channel, where it could be easily farmed, and could multiply without harming the surrounding crops and grassland. Eventually, however, like the musk-rat in our own day, the rabbit broke the bounds of its preserves and began to colonise the surrounding country. At first the abundance of predators, both birds and beasts, helped to keep it in check, but when other influences attacked the predators, as we shall see later, the rabbit was a *tertius gaudens*.

The influence of rabbits on vegetation has been ably summed up by Farrow, who shows that they cause heather heaths to degenerate into grass heaths, which rapidly revert to heather on being protected from rabbits ; that they are especially injurious to the taller growing plants and flower-spikes, some of which they entirely prevent from flowering ; that they greatly reduce the number of species of plant present in any area, especially the dicotyledons as against the grasses ; and that they limit tree-growth to the damper valleys. Thus in a rabbit-infested area, once trees have been destroyed by man for one purpose or another, rabbits will join with the domestic animals in preventing any young trees growing up again to replace them. As far as the London area is concerned, though there were rabbit warrens in Epping Forest and elsewhere during the Middle Ages, which supplied the citizens with rabbit meat and pelts, these catastrophic influences on vegetation probably did not begin to have serious effect till after the medieval period.

Another important introduction for food and sporting purposes was the pheasant, but since, whatever the original motive of the introduction, it has become the sporting bird *par excellence*, we will defer consideration of this also.

Medieval England was never very far from the edge of famine—there were several actual famines in London, such as that of 1314 when men ate dogs and made their bread of fern roots—and we may be sure that at all times small animals and birds were taken for the pot when opportunity offered. Hedgehogs are still eaten by gipsies, and must have formed tasty morsels at many medieval London suppers. Small birds as well often featured on the menu, as they still do on the continent, and we have the authority of a Tudor writer, Thomas Muffett, that the robin was then " esteemed a light and good meat." Since there were also many more birds of prey in the country round London in the Middle Ages, small song-birds must have been considerably less common than they are to-day in the rural parts of the London area.

Fish was a very important part of the medieval diet, and since communications were bad and fish does not keep, fish for Londoners meant freshwater fish, especially eels, though the Thames in those days was a famous salmon river. In Fitzstephen's day the Thames abounded with fish, and throughout the Middle Ages from Domesday onwards we find references to fish-weirs in the river.

In Domesday fish-weirs were recorded on the Thames in Middlesex at Fulham, Isleworth, Hampton, Shepperton and Staines, and on the Lea in Middlesex at Tottenham and Enfield. Fishponds or eel-ponds are also mentioned in several other places in the county ; at Harmondsworth, for instance, the Abbot of the Holy Trinity at Rouen held three mills valued at sixty shillings and 500 eels a year as well as fishponds valued at 1000 eels. At Stanwell, also near the confluence of the Colne with the Thames, four mills were valued at seventy shillings and 375 eels together with three weirs at 1000 eels, while Harefield, higher up the Colne, had four fishponds, calculated to produce 1000 eels a year for their fortunate owner, Richard the son of Earl Gilbert.

The Corporation of the City of London had a jurisdiction over the Thames, which stretched from Staines to a point near Yantlet Creek at the mouth of the Medway and lasted till the middle of the nineteenth century. Throughout the Middle Ages the City Fathers issued a series of ordinances (of which, however, scant notice seems to have been taken) for the removal of weirs that were either obstructing the navigation of the Thames or leading to the destruction of the brood fry. The removal of weirs was ordered, for instance, in the reign of Richard I, and, according to Stow, the Mayor caused all the weirs from Staines to the Medway to be removed in 1405. A close time for salmon was fixed by the Statute of Westminster in 1285, and under Richard II a hundred years later an Act was passed for the protection of fish-fry, especially that of salmon and lampreys. A statute of 1423 forbade the fastening of " nets and other engines called trinks and all other nets which be fastened continually day and night by a certain time of year to great posts, boats and anchors overthwart the river of Thames and other rivers of the realm," on the ground that they led to " as great and more destruction of the brood and fry of fish and disturbance of the common passage of vessels " as the weirs and kiddles. In 1349 three fishermen of Barking and Greenwich were convicted of catching " too small fish " on the east side of London Bridge.

In the year 1457 four " great fysshes " were caught in the Thames between London and Erith. They proved to be two whales, a sword-fish and a walrus. If, as seems possible, the " swordfish " was really a narwhal or sea-unicorn, all the " great fysshes " were in this case really mammals. All of them were also probably eaten, for the meat of whales, porpoises and seals was a regular feature at feasts at this time ; as " fysshes " they could legitimately be eaten on Fridays and during Lent. During Lent in 1246, for instance, Henry III ordered the Sheriffs of London to procure him a hundred pieces of the best whale, and two porpoises.

In addition to the fish caught in the rivers many were obtained from artificially stocked fishponds or stewponds, as for instance those on the south side of Old Palace Yard at Westminster, which were filled with pike and eels, and the " King's Pike Ponds " at Southwark. Old stewponds can still be seen in the north-west corner of Epsom Common. The word " stew " in this sense is derived from the Old French " estuier," to shut up.

Some contemporary lists and prices, quoted by Home, give a good idea of the kind of inroads made by medieval Londoners on the wild birds and fish of their neighbourhood. A list of the provisions presented to the young Edward III and his girl-queen Philippa in 1328 includes 24 swans at 5*s.* each, 24 herons and bitterns at 3*s.* 6*d.* each, four barrels of pickled sturgeon at £3 a barrel, and half a dozen barrels each of pike and eels at a total value of ten marks. Fifty years later a roast goose was sold for 7*d.*, a mallard for 4½*d.*, a pheasant for 13*d.*, a roast heron for 18*d.*, a roast bittern for 20*d.* and a roast curlew for 6½*d.* We may imagine that poulterers were not too particular about whether they actually served a curlew or something a bit smaller, such as a whimbrel or godwit snared or shot on migration. At this time, though, curlews and black-tailed godwits probably nested within sight of the City walls on the marshes south of the river. Smaller birds cost less, three roast thrushes fetching only twopence, five larks three-halfpence, and ten finches a penny. Here again, we may imagine that the identification of thrushes, larks and finches was not always very scrupulously carried out. After all, a roast starling looks much like a roast thrush and a roast pipit like a roast lark, so that this list by no means exhausts the number of different kinds of birds that found their way on to the platters of medieval Londoners.[1] Even so unusual a bird as the magpie appeared on a bill of fare drawn up for the use of the monks of Waltham Abbey, Essex, in 1059, along with the crane, thrush, partridge, pheasant, goose, fowl and falcon.

In 1418 it is recorded that oysters and mussels were sold in London at fourpence a bushel. It is not stated whether the oysters were good Whitstable or Colchester natives or merely dredged out of the Thames mud.

Mention has already been made of the gathering of herbs in the country round London, both for food and for medicinal purposes. Like Griselda in Chaucer's *Clerke's Tale,*—

> " And whan she homward came, she wolde bring
>   wortes and other herbes time oft,
>   the which she shred and sethe for hire living."

many London women must have earned their living in this way, or have replenished their own larders. A nutritious soup can be made from stinging nettles, which must have been abundant round medieval London ; sow-thistles (*Sonchus*), another common weed, also have a high nutritive value, which was taken advantage of by poor house-

---

[1] A cormorant was offered for sale at a Hampstead poulterer's in 1945, and moorhens and coots have also been seen in London shops in wartime.

Plate III

The great vine at Hampton Court   JOHN MARKHAM

Plate IV

The Regent's Canal as it passes through the Zoological Gardens   JOHN MARKHAM

wives at this period. There would also be wild fruits to be gathered, blackberries, raspberries, strawberries, elder-berries, hazel-nuts and sloes, while mushrooms and other fungi could be eaten by those bold enough to risk the fear of poison.

THE UTILISATION OF WILD ANIMALS AND PLANTS FOR CLOTHING AND OTHER DOMESTIC PURPOSES

In the prehistoric period man's needs for clothing, fuel, furniture and other domestic utensils must have resulted in a substantial slaughter of wild animals and destruction of woodland. By the Middle Ages, however, domestic animals were providing most of the skins and wool needed for clothing and much of the tallow needed for fuel, leaving only the continuing drain on the woodlands as a major influence under this heading.

Wood, and charcoal made from wood, was the chief fuel used in medieval London—" sea-coal " was only just coming in towards the end of the period—and the demand for fuel must have been an important cause of the destruction of woodland and clearance of scrub in the immediate neighbourhood of the City. To a smaller extent wood was needed for building houses and making furniture and various household utensils.

The right to cut firewood was an important one in the Middle Ages. The early court rolls of Wimbledon Common, for instance, have many references to the commoners' rights of cutting wood and furze. Until the nineteenth century many pollarded oaks, lopped annually, afforded winter fuel for the inhabitants, and though the pollards were cut down and sold in 1812 the tenants continued to be allowed to cut furze and dig peat. The rights of the commoners of Loughton to lop the hornbeams in Epping Forest (Plate 5) had the result of ultimately saving the forest for the public. (See p. 222.)

Londoners, however, at any rate after the twelfth century, had to look outside their own walls for firewood, and the Bishop of London had the right to levy one shilling on every cart of firewood that passed through Bishopsgate. As late as Tudor times the cutting of green boughs for sale in London as firewood was quite a trade at Enfield.

At the time of Domesday four Middlesex parishes were returned as having wood suitable for hedging or fencing ; these were St.

Pancras, Ossulston, Harlesden and Cranford. Though this wood was mainly required for agricultural purposes, doubtless some found its way into the City for fencing. Wood for making ploughs was an important requirement in those days, and it had its perquisites ; the woodward of Paddington, for instance, was entitled to the loppings from the timber felled for the lord's ploughs.

Even at the end of the Middle Ages carpets were still a great luxury, the ordinary man's house having rushes strewn over the floor. This had the advantage that there were no carpets to beat ; you merely swept the rushes out and strewed new ones. Rushes were very easy to obtain in medieval London, owing to the existence of large areas of marshy ground in Moorfields and along the Thames, and it is possible that the draining of these and consequent greater scarcity of rushes may have helped in the replacement of rushes by carpets as floor-coverings.

Though the skins of deer and other animals killed in the chase were used for clothing, and many of the poorer folk were clad almost exclusively in roughly tanned hides, this was probably not an important biotic factor. There must have been beavers in the London area at any rate in Saxon times, for the name Beverley (*beferithe*, beaver-stream) has come down to us attached to a brook that falls into the Thames opposite Fulham Palace, and forms part of the boundary between the counties of London and Surrey. Beavers were exterminated in England probably before the Conquest and mainly for their pelts. One of the uses of rabbit fur was as a substitute for beaver fur for caps, and it is perhaps significant that rabbits came in not long after beavers went out.

THE PROTECTION OF ANIMALS AS SCAVENGERS

Until the early part of the fourteenth century the streets of London were neither paved nor regularly cleansed, with results which may be imagined. One of the indirect consequences was the frequent occurrence of plagues and pestilences, as for instance in 961, and again in 1348, when the famous Black Death, now supposed to have been bubonic plague, carried off between a third and a half of the City's population. Even after ten years a third of the walled area of the City was still uninhabited.

Not only did all kinds of kitchen refuse, fæcal matter, carcases of dead dogs and the like lie openly about the streets, but the water that encompassed and flowed through the City was little better than a stagnant quagmire in the one case and an open sewer in the other. Medieval London must have been a paradise for mosquitoes, and it is not surprising to find that the ague, as malaria was then called, was prevalent. The Fleet ditch that washed the City's western walls was notoriously insanitary, and the Walbrook that ran through its centre equally so. Moorfields, the great stagnant fen that lay on the north side of the City wall, where Finsbury Square is to-day, made a splendid place for the youth of London to skate and slide when it was frozen over, but at other seasons the mosquitoes that bred in it must have slain their thousands. The Thames, judging by the amount of fish it held, must have been still relatively pure, in spite of the sewage poured into it from the Fleet and other outfalls.

In view of all this, it is not surprising to find that certain birds and beasts were protected as scavengers. In the earlier part of the Middle Ages, as we have seen, this function was performed by half-wild pigs, but as the built-up area became more extensive and standards of living slowly rose, the disadvantage of having droves of scraggy swine wandering about the streets and routing in everybody's kitchen midden became increasingly obvious, till in 1292 men were appointed to kill all stray swine. It was a different matter, however, with the kites and ravens. These were present in great abundance in medieval London, and were actively protected both by public opinion and by law for their good services.

Written records of the raven and the kite in London exist only for the end of the Middle Ages, but there is no reason to suppose that they were not useful scavengers throughout the period. According to Stow, there was such a murrain of kine in 1317 that even the dogs and ravens that fed on their bodies were poisoned. In the fifteenth century several writers refer to the large number of kites in London. When the King of Bohemia's brother-in-law, Baron Leo von Rozmital, visited London in 1465, his secretary, Schaschek, noted in his diary that nowhere had he ever seen so many kites as on London Bridge, adding that it was a capital offence to kill them. The same secretary to the Venetian Ambassador who marvelled at the sight of the Thames swans also noted in his journal about London :

" Nor do they dislike what we so much abominate, i.e., crows, rooks and jackdaws ; and the raven may croak at his pleasure, for no one cares for the omen ; there is even a penalty attached to destroying them, as they say that they keep the streets of the town free from all filth. It is the same case with the kites, which are so tame, that they often take out of the hands of little children, the bread smeared with butter, in the Flemish fashion, given to them by their mothers."

There are several similar statements relating to Tudor London. It is interesting to see here the earliest example of bird-protection in England for a reason other than sport.

HUNTING AND OTHER SPORT

The most picturesque of the relationships between man and his fellow animals in the Middle Ages, and certainly the aspect about which most has come down to us in the form of written records, is hunting and hawking. The City of London is intimately concerned, for it had hunting rights of its own over a wide area of what are now known as the Home Counties. Fitzstephen tells us that :

" Most of the citizens amuse themselves in sporting with merlins, hawks and other birds of a like kind, and also with dogs that hunt in the woods. The citizens have the right of hunting in Middlesex, Hertfordshire, all the Chilterns, and Kent, as far as the river Cray."

Some years earlier a charter of Henry I had granted the citizens of London " to have their chases to hunt as well and truly as their ancestors have had, that is to say in Chiltre, in Middlesex and in Surrey."

Sir Laurence Gomme held strongly that these hunting rights, coupled with the City's jurisdiction over the Thames from Staines to the Medway, were the remains of the *territorium* of Roman Londinium, and adduced them as an additional proof that London had survived as an intact community right through from the fifth century. As we saw in the previous chapter, the Royal Commission on Historical Monuments held this theory to be not proven, but saw no substantial

PLATE II

ERIC HOSKING

The famous maidenhair or ginkgo tree at Kew Gardens

PLATE 12

ERIC HOSKING

The nearest rookery to London, at Lee Green, S.E.12

reason why it should not be true. At all events it is attractive to imagine the proud Londoners maintaining their rights against successive waves of invaders, and winning back from the Conqueror's son the hunting privileges they had acquired under the Cæsars.

Middlesex has always stood in a very special relationship to London, the great city which has always been an enclave in England's second smallest county. Middlesex was never, as so many historians copying Stow have called it, a royal forest. The term " forest," which to-day is used loosely to mean extensive woodland, had in the Middle Ages a quite definite legal connotation. A forest was so called because it was land *foris* (without) the common law, and subject to a special law which aimed at preserving the king's hunting. A forest might thus contain much land that was neither wooded nor waste, though the extension of arable land in a forest was actively discouraged. Manwood, writing in 1598, defined a royal forest as follows :

" A forest is a certain territory of woody grounds and fruitful pastures, privileged for wild beasts and fowls of forest, chase and warren to rest, and abide there in the safe protection of the king, for his delight and pleasure ; which territory of ground so privileged is mered and bounded with unremovable marks, meres and boundaries, either known by matter of record or by prescription ; and also replenished with wild beasts of venery or chase, and with great covers of vert, for the succour of the said beasts there to abide : for the preservation and continuance of which said place, together with the vert and venison, there are particular officers, laws and privileges belonging to the same, requisite for that purpose, and proper only to a forest and to no other place."

William the Conqueror, as is well known, created many new royal forests in different parts of England, notably the barren tract of Hampshire that still goes by the name of the New Forest, and among the forests he took over from the Saxon kings was the ancient royal forest of Waltham, covering 60,000 acres in South-west Essex. To-day Epping Forest is one of the few remaining fragments of this great forest, where English kings from Harold to Charles I came to hunt, and where Henry VIII went to hear the signal that told of the execution of Anne Boleyn.

The forest laws were extremely harsh in their attempt to preserve enough deer for the king to hunt. In the words of the Anglo-Saxon chronicler in 1087 :

> William the Conqueror " made large forests for deer and enacted laws therewith, so that whoever killed a hart or a hind should be blinded. As he forbade the killing of deer, so also the boars. And he loved the tall stags as if he were their father. He also appointed concerning the hares that they should go free. The rich complained and the poor murmured but he was so sturdy that he recked nought of them."

In the Forest of Waltham no fences might be maintained high enough to keep out a doe with her fawn, nor were the farmers allowed to drive the deer from their crops, on which they fattened. Buildings might not be erected without the permission of the Forest authorities, " because of the increase of men and dogs and other things which might frighten the deer from their food." The consent of the authorities was also needed to cut down trees, while the Crown claimed the right to enter the woods of any private owner within the range of the forest and cut the branches of trees as " broust " for the deer's winter feed. This last right was exercised as late as the early nineteenth century. Dogs in the district had to be " expeditated " or " lawed," which meant that three claws of their forefeet were cut close to the ball of the foot to prevent them chasing the deer.

Dogs might incur serious penalties for their owners under the Forest Charter, which was extorted from King John as a kind of appendix to Magna Charta, and replaced an earlier and even harsher charter of Canute :

> " If a greedy ravening dog shall bite a wild beast, then the owner of the same dog shall yield a recompense to the king for the same according to the value of a freeman which is twelve times a hundred shillings. If a Royal beast shall be bitten, then the owner of a dog shall be guilty of the greatest offence."

The Forest Charter also laid down penalties for men chasing the deer :

> " If any freeman shall chase away a deer, or a wild beast out of

the Forest : whether the same be done by chance, or of a set purpose, so that thereby the wild beast is forced by swift running to loll out the tongue, or to breathe with his tongue out of his mouth : he shall pay to the king ten shillings amends for the same offence : but if he be a servile person, then he shall double the same recompense : but if he be a bondman, then he shall lose his skin."

Buxton quotes a quaint old rhyming charter of the appointment of one of the officers of the Forest by Edward the Confessor, which gives an idea of the fauna of the Forest at that time :

> " Ich, Edward Koning,
>   Have yeven of my forest the keping
>   Of the hundred of Chelmer and Dancing
>   to Randolph Peperking and his kindling
>   Wyth heorte and hynde, doe and bocke,
>   Hare and foxe, Catte and brocke,
>   Wild fowel with his flocke
>   Partrich, fesant hen and fesant cocke,
>   With green and wylde, stob and stocke
>   to kepen and to yeinen, by al her might,
>   Both by day and eke by night.
>   And hounds for to hold
>   Good swift and bold
>   Four greyhounds and six racches
>   For hare and foxe and wilde cattes ; "

Already there were both red deer (hart and hind) and fallow deer (doe and buck) in the Forest, together with hares, foxes, wild cats, badgers, partridges and pheasants. Of all these only the wild cats have not survived till the twentieth century. The identity of the " wild fowel with his flocke " is partly revealed by Chapter 13 of the Forest Charter, which reads : " Every Freeman shall have, within his own woods, ayries of Hawks, Sparrow-Hawks, Faulcons, Eagles and Herons ; and shall also have the Honey that is found within his woods." The hawks and falcons here referred to were probably goshawks and peregrine falcons.

The true beasts of the forest in the Middle Ages were the three kinds of deer, red, fallow and roe, together with the wild boar. These were also beasts of the chase, a chase being the great landowner's equivalent of the royal forest, where he had special hunting privileges, though not completely without the law. Lower down the scale came the warrens, where there were exclusive rights of the hunting and taking of certain other game animals, notably hares, rabbits, foxes, pheasants, partridges, woodcock and herons.

There was at one time a royal warren at Staines, which was disafforested (i.e., the special game rights were rescinded, not the woodland uprooted) in the reign of Henry III. Special game laws also applied in warrens ; for instance, dogs in royal warrens had to be lawed, and other lords of warren had the power of impounding dogs, snares and traps.

Finally, the general public had the right to hunt wild animals in any unenclosed lands outside the forest limits, and where no rights of warren applied. Within the Home Counties area in which Londoners had their special rights this would naturally apply only to citizens. Taking a couple of dogs and a bow out for an afternoon's hunting in Southwark Woods must have been a typical recreation for many medieval London citizens. For the more substantial citizens there was the chance of a day out with the Lord Mayor's staghounds, which are heard of at least as early as the fifteenth century.

In a few places round London, notably Ruislip and Enfield, noblemen kept parks or enclosures for various kinds of wild beasts and game. These differed from chases chiefly in being fenced in.

Hawking was in high favour as a sport throughout the Middle Ages, and as we have seen was indulged in by the citizens of London. In the fourteenth century Richard II kept his hawks in the mews at Charing Cross. The importance of the birds of prey mentioned in the Epping Forest charter lay in the value of the eyries, which provided young birds for training for falconry. The Chronicle of Roger of Wendover contains the following anecdote of hawking near London :

" In the same year [1191] a young man of the bishop of London's household taught a hawk (*nisum*) especially to hunt teals , and once at the sound of the instrument called a tabor by those who dwelt on the river's bank, a teal suddenly flew quickly away ; but the hawk baffled of his booty intercepted a pike swimming in the

Plate V

A horse-chestnut in bloom by the Grand Union Canal in Cassiobury Park, Watford, Herts.

Plate VI

A brown rat in a London warehouse   JOHN MARKHAM

Nests of house-mice in flower-pots in a greenhouse   JOHN MARKHAM

water, seized him, and carried him apparently forty feet on dry land."

It has been suggested that the hawk might have been an osprey, though the word *nisus* usually refers to a sparrow-hawk.

It is probable that hunting did not affect the status of most of the beasts and birds that were hunted in the London area in the Middle Ages. The exceptions were the two largest beasts, the wild bulls mentioned by Matthew Paris and Fitzstephen, and the wild boars, the first of which appear to have completely disappeared well before the end of the fifteenth century, while the second did not survive the following century in the London area. Deer must still have been abundant in the extensive woodlands of Middlesex, West Essex and Surrey.

Among game birds the pheasant was a newcomer which we probably owe to the Normans. Much has been made of the fact that one pheasant was regarded as equivalent to two partridges in the bill of fare at Waltham Abbey in 1059, quoted above. While this indicates that the pheasant was known in England immediately before the Conquest, as does also the rhyming charter of Edward the Confessor, it must be remembered that Norman influence was strong in England throughout the Confessor's reign, which began in 1042, and the birds may well have been introduced some time in the half century before the Conquest. It is also possible that the introduction had been made only in the Forest of Waltham and one or two other places up to 1066. Admittedly, remains of pheasants have been found at Roman settlements in Britain, but while this points to the introduction of the pheasant by Roman epicures, it does not prove its survival in a wild state throughout the six troubled centuries between the departure of the legions and the infiltration of the Normans. Indeed, the greater number of birds and beasts of prey then present in the English woodlands make it unlikely that a bird which has to be rigorously protected to enable it to survive in the much more favourable conditions of to-day could have lasted unprotected for 600 years.

OTHER HUMAN INFLUENCES ON ANIMAL AND PLANT LIFE

In the Middle Ages pleasure and luxury on a scale big enough to influence the status of animals and plants was confined to the rich, and they were so few that the total effect was negligible. They had their gardens, no doubt, with trees and bushes and flower-beds that provided the beginnings of a habitat for the garden association of birds and insects and plants that we know to-day. They also kept pets, and in their houses you might come across one of those queer animals called cats, for the tale of Dick Whittington shows that in the late fourteenth century cats were regarded as rare and luck-bringing beasts. The lucky qualities that to-day inhere only in black cats probably belonged then to all cats. Two favourite birds in the mansions of the nobility are said to have been the parrot[1] and the magpie, the latter of which also earned its keep in the poultry yard in much the same way as the sacred geese in Rome.

The common people had few recorded customs that are likely materially to have affected wild birds, beasts or plants. Probably the boys of London pursued the barbarous practice of wren-baiting, which consisted in beating the hedgerows till a terrified wren flew out and was killed. Certainly the young people of London would gather great armfuls of white may-blossom from the hawthorns to decorate the maypole on May Day. Old May Day fell about the present May 10, when the thorn-trees are more regularly in bloom than on the present May 1.

The rudimentary zoo at the Tower of London, which was the origin of the phrase " seeing the lions in the Tower," does not strictly belong to a book on natural history—except insofar as the polar bears were let out into the river on collars and chains to catch salmon—but it is of sufficient interest to be mentioned here. Wild beasts have been kept in the Tower almost ever since it was built, though most of them were transferred early in the nineteenth century to become the nucleus of the present zoo in Regent's Park. Their last survivors are the two ravens. Loftie's description of the early history of the menagerie is worth quoting in full :

" Henry I had a collection of lions, leopards and other strange animals. Three leopards, in allusion perhaps to the royal heraldry, were presented to Henry III by the Emperor Frederick II. This king indulged his zoological tastes at the expense of the City, whose

[1] Ray and Willughby in their *Ornithology* quoted Gesner's story of Henry VIII's parrot, which is said to have cried out, on falling into the Thames from a window in the Palace of Westminster, " a boat, a boat for twenty pounds."

greatest oppressor he seems to have been in so many other respects. The sheriffs had to arrange in 1252 for the safe-keeping of a white bear from Norway. They ' provided four pence daily, with a muzzle and iron chain, to keep him when *extra aquam* and a stout cord to hold him when a-fishing in the Thames.' Two years later an Elephant arrived from France [the first since the Romans left]. He landed at Sandwich and the sheriffs had to provide for him ' a strong and suitable house,' and to support him and his keeper. ' At the time when the allowance for an esquire was one penny a day,' remarks Mr. Clark, ' a lion had a quarter of mutton, and three halfpence for the keeper ; and afterwards sixpence was the lion's allowance ; the same for a leopard, and three halfpence for the keeper.' "

One of the modern equivalents of " seeing the lions in the Tower " is feeding the pigeons at St. Paul's and in Trafalgar Square. (Plate 7.) The colonies of semi-wild pigeons in London, all doubtless descended from birds that escaped from medieval dovecots, are of very ancient origin, and were already well established at the end of the fourteenth century. Robert de Braybrooke, Bishop of London, complained in 1385 that :

" There are those who, instigated by a malignant spirit, are busy to injure more than to profit, and throw from a distance and hurl stones, arrows and various kinds of darts at the crows, pigeons and other kinds of birds building their nests and sitting on the walls and openings of the church, and in doing so break the glass windows and stone images of the said church."

One suspects that the " malignant " persons were just ordinary small boys, and that the lament reflects the unchangingness of human nature. One would dearly like to know what were the other birds that nested on old St. Paul's. Kites ? Storks ? Owls ? Jackdaws ? or just our old Cockney friend Philip Sparrow ?

A more malignant creature than those who broke the windows of St. Paul's was the black rat, which arrived uninvited, unwanted and unsung as a stowaway some time in the thirteenth century. Barrett-Hamilton and Hinton consider that the arrival of the black rat was one of the less desirable consequences of the Crusades, since it probably came here direct from Palestine in the ships of our crusading navies. The well-known story of the Pied Piper of Hamelin shows what a pest the black rat rapidly became on the Continent, and even in England it was sufficiently common by the fourteenth century to evoke demands on the apothecaries for rat-poison. In Chaucer's words :

" And forth he goth, no lenger wold he tary,
    Into the toun unto a Pothecary,
    And praied him that he him wolde sell
    Som poison that he might his ratouns quell .

References for Chapter 4

Barrett-Hamilton & Hinton (1910-21), Bayes (1944), Buxton (1901), Darby (1936), Domesday Book (1086), Eversley (1910), Farrow (1917), Fitzstephen (12th cent.), Glegg (1935), Gomme (1914), Gurney (1921), Harting (1880), Home (1927), Loftie (1883), Page (1923), Perceval (1909), Pigott (1902), Ritchie (1920), Stow (1598), Ticehurst (1928, 1934, 1941), Vulliamy (1930).

THE WEN BEGINS TO SWELL

" But what is to be the fate of the great wen of all ? The monster, called . . . ' the metropolis of the empire ' ? "

WILLIAM COBBETT, *Rural Rides*, 1821.

FOR London the Middle Ages end somewhere between the accession of Henry Tudor in 1485, which brought with it an immediate expansion of prosperity and population, and the dissolution of the monasteries by his son some fifty years later. The two main outward signs of the transition were the change in building materials, summed up in James I's *mot* " from sticks to bricks," and the disappearance from the streets of the innumerable religious folk, monks, friars and nuns, with their semi-immunity from the law of the land, and their vast establishments occupying valuable space within the City walls. The expansion of the City after the troubled times of the Wars of the Roses began as soon as the strong Tudor government got under way, and was only slightly delayed by the building over of the monastic lands within the existing built-up area after 1536.

The wen, as Cobbett loved to call London, has been expanding ever since. Repeated but completely unavailing attempts to stem the inexorable advance of London over the countryside of Middlesex and Surrey have been made, from the City Corporation's complaint to the Privy Council in 1580 " of the vast increase of new buildings and number of inhabitants within the City and suburbs of London . . . which was imagined would prove of dangerous consequence, not only to this great metropolis, but likewise to the Nation in general if not timely remedied," to the Greater London Plan of 1944, with its proposals for an extensive Green Belt to cut Outer London off from the rapidly developing dormitory towns twenty miles or more from the centre. But London, like Ol' Man River, " jus' goes rollin' along " ; at least it did so up to 1939.

It is not easy to fix a dividing line between the time when London was still a manageable entity and the present amorphous sprawl over five counties, but perhaps 1851, the year of the Great Exhibition, that stupendous Victorian landmark, is as good as any other. Besides neatly cutting the nineteenth century in half, it is only four years before the foundation of the Metropolitan Board of Works, forerunner of the London County Council, whose area coincided almost exactly with the present County of London, which thus marks the approximate extent of the built-up area in the middle of the nineteenth century. The fixing of this line also gives us a period of almost a hundred years in which to trace the full burgeoning of the Wen until now it stretches almost unbroken from Hertford in the north to Reigate in the south, and from Tilbury and Gravesend in the east to Staines in the west.

In early Tudor times London (Map 3) still hardly covered more than the present City area, with a penumbra of suburbs up to half a mile all round the walls, and narrow strips of riverside development upstream to Westminster, downstream to Wapping, and on the south bank from Blackfriars to Bermondsey. By the time of the Great Exhibition London had linked up with Chelsea, Paddington, Camden Town, Islington, Bethnal Green and Poplar in the north, and with Greenwich, Camberwell, Brixton and Battersea in the south. Already it was a very far cry from the days when Colet was Vicar of Stepney, and More wrote to him (1504) :

" Wheresoever you look, the earth yieldeth you a pleasant prospect ; the temperature of the air fresheth you, and the very bounds of the heavens do delight you. Here you find nothing but bounteous gifts to nature and saint-like tokens of innocency."

The population of London throughout the Middle Ages had probably been somewhat below its Roman prime of 50,000. Creighton, the epidemiologist, reckoned that between the reigns of Richard I and Henry VII the population of London fluctuated between 40,000 and 50,000, but that with the Tudors a continuous expansion began, which by 1532 had increased the population to about 62,000. A generation later he put the number of Londoners at 93,000, and by the end of the century, 1593, at 140,000 to 150,000, rather more than half of them already living without the City area in the liberties and suburbs.

By the year of the Great Plague, according to Bell, London contained not far short of half a million people, of whom some 100,000 died in the epidemic. By the end of the seventeenth century the total population of London appears to have reached nearly 675,000, of whom only 139,000 lived in the old City. It was not for another hundred years, in 1801, that the first actual census was taken, when London proved to have nearly 900,000 inhabitants, more than ten times the population of its nearest rival in England, Manchester-Salford. Already there were 123,000 people living in the five outer parishes of Kensington, Chelsea, St. Marylebone, Paddington and St. Pancras, which a hundred years before had been semi-rural villages with fewer than 10,000 inhabitants between them. By the year of the Great Exhibition, the population of the present area of the County of London had risen to 2,363,341.

THE EFFECTS OF BUILDING

From the close of the fifteenth century onwards far and away the most important influence of London on the natural animal and plant communities that shared the lower Thames valley with the rapidly expanding human community was that of the spread of streets and buildings. The suppression of the religious houses handed over to the builder many of the islands of open space that had been preserved, like the Inns of Court to-day, either within or immediately outside the City walls. This flinging of a victim to the wolves, however, hardly slowed up the pursuit, and certainly its influence did not last beyond 1570.

The dissolution of the monasteries had another important consequence, the transformation of the abbots' and canons' farms near London into the country mansions of wealthy aldermen, surrounded by extensive parks. This began the familiar sequence of " developing " London's countryside. First, the farmland was enclosed in parks, either for deer or for mere pleasances ; then the parks were broken up into smaller plots on which were erected the large villas of the more prosperous citizens ; and finally the tide of closely built streets and lanes overtook the district. The park stage was sometimes missed out and the farmland converted straight into good-class residential property, but the stage of large houses with gardens has usually preceded that of continuous built-up area, particularly during the period up to the Great Exhibition. An example of Tudor development from Stow is

quoted later on, where a market garden in Houndsditch gave way to
" many fair houses of pleasure."

A glance at Maps 3 and 4 shows the successive stages of London's
spread. During the 16th and 17th centuries expansion was comparatively slow, being partly held up by the aftermath of the Great
Fire of 1666. Even by 1745 the continuous built-up area only extended
as far as Lambeth Bridge and Park Lane in the west and to Shoreditch
and Mile End in the east, while in the north very little was built up
beyond Oxford Street, and the southern bridgehead had grown by
less than half a mile all round. In the next 75 years almost as
much development took place as in the preceding two hundred. By
1820 the advancing tide of London was lapping up to Vauxhall Bridge,
Victoria, Edgware Road, Regent's Park, City Road, Limehouse,
Rotherhithe and Lambeth. Then in the small space of the next sixteen
years an extensive area was added to the fringes of London, and
nearly all the gaps in the inner core were built up. In the years
following Waterloo, London surged over Belgravia, Paddington, St.
Pancras, Islington, Hoxton, Bethnal Green, Bow Road, Poplar, Deptford, Kennington and Walworth. This was the period of the most
rapid development of inner East and South London. In another
thirty years London would be found taking on some of its modern
aspect. By 1872 continuous tongues of building stretched from the
central core right out to Walham Green, Hammersmith, Kensal
Green, Hampstead, Highgate, Finsbury Park, Stamford Hill, Clapton,
Hackney, Old Ford, East India Dock, Blackheath, New Cross,
Peckham, Norwood, Streatham and Tooting. There were still large
patches of green country interlocking with the tongues of buildings,
as near the centre as South Kensington, Royal Oak, Primrose Hill,
Barnsbury, Hackney, Burdett Road, Rotherhithe, South Bermondsey,
Brixton and Battersea. If only the growth of London could have been
controlled in 1872, half the work of the County of London and Greater
London Plans would not have needed to be done.

The country over which this flood of buildings spread was to a
very large extent agricultural, some of it the richest market garden
land in England, but there was also some marshland in the valleys
of London's many little streams, the Fleet, the Tyburn, the Westbourne, and on the flat expanse of South London, where pools lay
stagnant from the Conquest to Waterloo. When the Isle of Dogs was
built over, some of the finest sheep-grazing pasture in the country,

Plate VII

Male and female cockroaches with egg-sac   s. BEAUFOY

Spiders' webs   s. BEAUFOY

Plate VIII

A cedar tree in Highgate cemetery   JOHN MARKHAM

rivalling Romney Marsh, was lost. When London's tentacles embraced Chelsea and Fulham, the metropolis sacrificed one of its principal sources of fruit and vegetables.

Indeed, none of the land that has been engulfed by London's bricks and macadam since the Romans turned the first virgin sods has been snatched straight from a primeval state, save perhaps a few marshy floors. When land near London has come to be finally built over, it has almost invariably been either farmland, including market gardens, or the parkland or pleasure gardens surrounding a mansion or large villa. The most notable exception was the development of the site of modern Belgravia by Thomas Cubitt. Here, as his great-grandson, Sir Stephen Tallents, tells us, in 1824 Cubitt found 140 acres of waterlogged swamp known as the Five Fields, between Hyde Park Corner, Sloane Street and Grosvenor Place, with the small River Westbourne flowing through it just to the east of Sloane Square. Within five years he had replaced the swamp by Belgrave, Eaton and Lowndes Squares and Chesham and Eaton Places. He then turned his attention to the area between Eaton Place and the Thames, stripped off all the clay topsoil to make bricks, and both replaced this soil and filled in the reservoirs of the Chelsea Waterworks with the earth from Telford's excavations for St. Katherine's Dock. Thus South Belgravia and Pimlico actually stand on soil, including the contents of a church-yard, from a parish on the other side of London, St. Katherine by the Tower.

The marshiness of much of the West End of London before it was built over is reflected in the many records and reminiscences of the occurrence of woodcock and snipe there a century or more ago. General Oglethorpe, who died in 1785, often shot woodcock on the site of Conduit Street, and William Wilberforce claimed in 1815 to have spoken with a man who had shot woodcock in the parish of St. Martin-in-the-Fields. In 1879 there were said to be men still living who had shot snipe in the Five Fields, and Sir Charles Lucas was able to tell the British Association in 1914 that he had heard a lady relate that her grandfather used to shoot snipe in the same locality. As Cubitt did not reclaim the Five Fields till the middle of the third decade of the nineteenth century, it would have been perfectly possible for boys who had shot snipe there to survive till the early years of the present century.

Across the river extensive marshes existed until quite a late date

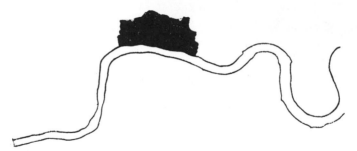

1560   After Laurence Gomme (Journal of the R.G.S., 1912)

1668   After Laurence Gomme (Journal of the R.G.S., 1912)

1745   After John Rocque : London, 1741-1745

1820 After Cary's New Plan of London, 1820

1836 After Cary's New Plan of London, 1836
MAP 3 The Spread of London, 1560-1836
The black outlines show the built-up area of London in 1560,
1668, 1745, 1820 and 1836

in the present boroughs of Bermondsey, Southwark and Lambeth. Lambeth Marsh really was a marsh, and the Effra river flowed openly to the Thames instead of through a sewer. William Curtis, writing at the end of the eighteenth century, noted the rare dock *Rumex maritimus* as growing near St. George's Fields, where St. George's Circus now stands, and remarked of the handsome aquatic plant, the flowering rush (*Butomus umbellatus*), "a few years since it was found growing in St. George's Fields ; but the improvements making in that, and other parts adjacent to London, now oblige us to go farther in search of this, and many other curious plants. About the Isle of St. Helena, near Deptford, and in the marshes by Blackwall, it is found in great abundance." The marsh valerian (*Valeriana dioica*) was very common in the meadows and osier grounds about Battersea, and the buckbean (*Menyanthes trifoliata*) was abundant in the Battersea meadows and around the Isle of St. Helena towards Rotherhithe. Curtis also recorded the buckbean as common in the marshes about Staines, in many of which it was the principal plant, while as late as 1869 Trimen and Dyer described it as abundant in the bog behind Jack Straw's Castle on Hampstead Heath, where it was first recorded by Thomas Johnson in 1629. By 1913 the buckbean had sadly to be recorded as extinct in this historic locality and the nearest recent locality to London is Wimbledon Common.

G. Graves, who wrote about British birds during the Waterloo decade, recorded all three species of harrier from the South London marshes. The hen-harrier (*Circus cyaneus*) was apparently no uncommon sight, skimming over the fields known as Rolls's Meadows by the side of the Kent Road. In May, 1812, a pair of Montagu's harriers (*Circus pygargus*), evidently nesting near by, were shot in Battersea Fields, while in the same year a marsh-harrier's (*Circus æruginosus*) nest was found by an osier pond near the Grand Surrey Canal near Deptford. The nest was made of sticks, grass and leaves and decayed stalks of the flowering rush, and was situated on a small hillock just above the water's edge. To add to the similarity of the South London marshes in the 1810's to the Norfolk Broads to-day, the bearded tit (*Panurus biarmicus*) seems to have been a common bird along the whole length of the Thames from Oxford to the estuary. Graves shot bearded tits by the side of the Surrey Canal, on Sydenham Common, and in Blue Anchor Lane, which leads from Bermondsey to Deptford. Other marsh birds that frequented the outskirts of London at this time, and

Allotments on Barking Level, Essex ERIC HOSKING

Backyard pig farming ERIC HOSKING

PLATE 14

Cock Blackbird at nest IAN M. THOMSON

had presumably done so for hundreds of years previously, were the common snipe, which nested abundantly in osier beds along the Surrey Canal, and the spotted crake (*Porzana porzana*), to-day one of the rarest of British breeding birds, which was at that time said to occur in greater abundance within a few miles of London than anywhere else in the British Isles. Graves knew it to breed in Rolls's Meadows, where he used to see the hen-harriers.

The Thames in London was still unembanked throughout this period, and many water birds and plants flourished along its muddy shores right into the centre of London. The arrowhead (*Sagittaria sagittifolia*), which was first recorded as a British plant by Lobel in the Tower ditch in 1570, could be seen at the York Water Gate, where the Victoria Embankment Gardens are to-day, as late as 1812, and in 1833 several pairs of reed-warblers nested annually in the reeds on the river bank between Battersea Bridge and the Red House Tavern in Battersea Fields. As late as the middle of the nineteenth century the rare sedge *Scirpus triqueter* was common on the Surrey shore of the Thames between Battersea and Richmond ; Petiver had found it near the Horseferry at Westminster in the late seventeenth century ; even within recent years it has been found at Barnes, Mortlake and Kew.

An excellent way of tracing the growth of London is to look back at the plants found by the early botanists in their wanderings on the outskirts of the City. So many botanists have lived in London and left records of their work—Turner, Penny, Lobel, Gerard, De L'Ecluse, Parkinson, Johnson, Tradescant, Merret, Morison, to mention only those who lived up to 1700—that it is quite impossible to refer to them all in a work of this scope. The few records cited here barely touch the fringe of the rich field that awaits the thorough historian of London's flora.

William Turner, "the Father of British Botany," is the first botanist who has left us records of plants growing in or near London. In his *Herball* and other botanical writings published between 1548 and 1568 he recorded sixteen plants from Middlesex. These included the pepperwort (*Lepidium campestre*), recorded from two London gardens, where it was perhaps cultivated ; the square-stalked St. John's Wort (*Hypericum acutum*) in Syon Park, Isleworth, in 1548 ; dropwort (*Filipendula hexapetala*), in " great plenty . . . in a field adjoining to Sion House . . . on the side of a meadow called Sion Meadow " ; great burnet (*Sanguisorba officinalis*), " muche about Sion " ; chamo-

mile (*Anthemis nobilis*) " in mooste plenty of al " on Hounslow Heath ;
the rare cotton thistle (*Onopordum acanthium*) at Syon ; the thyme-leaved
speedwell (*Veronica serpyllifolia*) " in a parke besyde London " and " in
divers woddes not far from Syon " ; pennyroyal (*Mentha pulegium*) in
a wet place on Hounslow Heath ; great wild basil (*Clinopodium vulgare*)
and spurge laurel (*Daphne laureola*) at Syon ; annual mercury (*Mercurialis annua*), which " beginneth now to be knowen in London, and
in gentlemennis places not far from London " ; and the lady's tresses
orchid (*Spiranthes autumnalis*) and the bluebell (*Scilla non-scripta*), both
at Syon, the latter abundantly. The records for all these plants,
together with *Sherardia arvensis*, goat's-beard (*Tragopogon pratensis*) and
wood-spurge (*Euphorbia amygdaloides*) are the first not only for Middlesex
but for Britain also. The predominance of records from Syon House
and Park is due to the fact that Turner was physician to Lord Protector
Somerset, whose seat it was.

John Gerard, whose *Herball* was published in 1597, was more of a
Londoner than Turner, who indeed spent very little of his time in the
city. Gerard lived in Holborn, and in the course of his frequent
botanical rambles in the country between London and Hampstead,
added seventy-three species of plants to the Middlesex list, including
the alder buckthorn (*Rhamnus frangula*), wild service tree (*Sorbus torminalis*), golden rod (*Solidago virgaurea*), bell-heather and cross-leaved
heath (*Erica cinerea, E. tetralix*), whortleberry (*Vaccinium myrtillus*) and lily
of the valley (*Convallaria majalis*), the last-named of which grew in great
abundance on Hampstead Heath. The Heath was Gerard's favourite
hunting ground, as it was of so many later London botanists, and he
described the hills on which it stands as " drie mountains which are
hungrie and barren." Here he found, among other things, the butcher's
broom (*Ruscus aculeatus*), the cow-wheat (*Melampyrum pratense*), which
grew " in all parts among the juniper and bilberrie bushes," cotton-
grass (*Eriophorum*), which was still abundant in the bog behind Jack
Straw's Castle in the 1860's, golden-rod and tutsan (*Hypericum androsæmum*), which was common in the drier parts of the woods round
Hampstead and Highgate till many of them were grubbed up and
cultivated. Pennyroyal he found growing " on the common neere
London called Miles ende ; . . . whence poore women bring plentie
to sell in London markets, and sundrie other commons neere London."
Clary (*Salvia verbenaca*) grew wild " in divers barren places . . .
especially in the fields of Holburne near unto Graies Inn."

Southwards, too, Gerard often went botanising, sometimes accompanied by Matthias de Lobel, whose name has come down to us attached to a genus of plants, one of which is a popular garden plant often used with white alyssum and red geraniums to give a patriotic red, white and blue effect. Water crowfoot (*Ranunculus aquatilis*) he found growing in a ditch near the Earl of Sussex's garden wall in Bermondsey Street, while water hemlock (*Cicuta virosa*) flourished in the ditches by a causeway on the way to Deptford. The pinks that still go by the name of Deptford pink (*Dianthus armeria*) were to be found abundantly " in our pastures neare about London, especially in the great field next to Detford by the path side as you go from Redriffe (Rotherhithe) to Greenwich." In the following century Parkinson noted these pinks " among the thicke grasse towards Totnam Court, near London," and Newton " in a little wood cut down on the right hand side of the road, a little beyond the bottom of the hill beyond Highgate. Nobody has recorded the Deptford pink from Middlesex since about 1680, and it has long vanished from the south side also.

The walls of Elizabethan London seem to have been particularly rich in flora. The great water radish (*Rorippa amphibia*) grew " in the joints or chinks among the mortar of a stone wall that bordereth upon the river Thames by the Savoy in London, the which you cannot finde but when the tide is much spent " ; the wall rocket (*Diplotaxis tenuifolia*) was abundant, most brick and stone walls about London being covered with it ; it still grew on the front of a house in the Westminster School yard in 1867. Rue-leaved saxifrage (*Saxifraga tridactylites*) was found on brick and stone walls and old tiled houses, " which are growen to have much mosse upon them and upon some shadowie and drie muddie walls. It groweth plentifully upon the bricke wall in Chauncerie Lane belonging to the Earl of Southampton, in the suburbs of London." Whitlow grass (*Erophila verna*) grew on the same wall.

The next London botanist of importance, and one of particular significance in the history of the flora of London, was Thomas Johnson, aptly described by Trimen and Dyer as an " indefatigable botanist and excellent man." Johnson was a leading member of the Society of Apothecaries, who were in the habit of going on botanical rambles near London. His *Iter Plantarum Investigationis Ergo Susceptum a Decem Sociis in Agrum Cantianum Anno Dom. 1629, Julii 13* is an account of one of these expeditions undertaken in Kent by Johnson and nine companions on July 13, 1629 ; it is also the first printed account of a botanical excursion in England.   In an appendix to his little book Johnson gave in *Ericetum Hamstedianum seu Plantarum ibi crescentium observatio habita Anno eodem 1 Augusti* a description of a similar visit to Hampstead Heath on August 1, 1629, in which seventy-two plants are listed.   This constitutes the first published account of a botanical excursion in the London area.   Three years later Johnson published a list of ninety-seven plants growing on Hampstead Heath, which together with the previous list may claim to be the first local flora printed in England.   These two lists are reproduced in Appendix A. Many of the plants he mentioned, and he omitted some of the commonest, may still be found on or near the Heath.   These include tormentil (*Potentilla erecta*), angelica (*Angelica sylvestris*), lesser spearwort (*Ranunculus flammula*), and marsh marigold (*Caltha palustris*). Others, such as the lilies of the valley also noted by Gerard, have long gone.

About 150 years later William Curtis brought out his great *Flora Londinensis*, which began as a description of " such plants as grow wild in the environs of London," but later covered a much wider field. Among the unusual plants he recorded in places which within the next fifty years or so were to be covered with bricks and mortar were the water violet (*Hottonia palustris*), which " among a variety of other places " might " be found in a ditch on the right hand side of the field way leading from Kent-street Road to Peckham " ;   blinks (*Montia fontana*), which grew on Blackheath and Hampstead Heath ;   the periwinkle (*Vinca minor*), which flourished in the hedge of a field in Lordship Lane, Dulwich ;   the danewort or dwarf elder (*Sambucus ebulus*), now extremely rare round London, which then grew on Lambeth marsh and at Upton, Essex ;   the handsome fritillary (*Fritillaria meleagris*), which occurred " in meadows betwixt Mortlake and Kew " and also at Ruislip Common and Enfield ;   several stonecrops, *Sedum album* scarce, but on a chapel wall in Kentish Town and on a wall at Bromley-by-Bow, *S. acre* very common on houses, walls and gravelly banks, and *S. dasyphyllum* frequent in such places as the Chelsea Hospital, Kensington gravel pits and Acton Road ;   the corn-cockle (*Lychnis githago*) very common in most cornfields ;   and succory (*Cichorium intybus*) in Battersea Fields, " which exhibit bad husbandry in perfection."   As in Gerard's day the rocket was very plentiful on walls, especially at the Tower, Bedlam and Hyde Park.   The ivy-leaved

Plate IX

The Chelsea Physic Garden JOHN MARKHAM

Plate X

A typical Inner London street with no gardens JOHN MARKHAM

Garden of a blitzed suburban house overgrown with rose-bay willow-herb and other weeds
JOHN MARKHAM

toadflax (*Linaria cymbalaria*) was another common wall-plant, plentiful on walls at the Chelsea Physic Garden (whence this Mediterranean plant is believed to have originated in Britain) at the Temple and by the stream under the Vauxhall turnpike. Yet another wall-plant, pellitory of the wall (*Parietaria officinalis*) was common among rubbish and on walls, especially by the Thames above and below Westminster Bridge ; in 1945 it can still be seen growing in profusion on a fragment of the Roman wall.

Trimen and Dyer's *Flora of Middlesex*, one of the first comprehensive county floras, which appeared in 1869, contains many records of the plants that had been eliminated in the spread of London in the previous two generations. Traveller's joy or old man's beard (*Clematis vitalba*), for instance, could be found in the Edgware Road and on the way from Chelsea to Fulham as late as 1830. The stroller in Marylebone Fields in 1817 would have discovered the wood anemone (*Anemone nemorosa*), lesser celandine (*Ranunculus ficaria*), white campion (*Lychnis alba*) and hogweed (*Heracleum sphondylium*). Eelbrook Common, near Walham Green, lost its marsh plants when it was drained ; it is hard to believe nowadays that marsh marigolds once grew there. The rosebay willow-herb (*Epilobium angustifolium*), on the other hand, to-day one of the commonest flowering plants in the County of London, was both in Curtis's day and when Trimen and Dyer wrote, a rare plant. A catalogue of plants collected in Islington in 1837-42 gives the wallflower (*Cheiranthus cheiri*), wall-pepper (*Sedum acre*), and rue-leaved saxifrage (*Saxifraga tridactylites*), as growing on walls in Canonbury and Islington, and watercress (*Nasturtium officinale*) on the site of Canonbury Square.

Right up to the beginning of the nineteenth century, the country was never very far from the centre of London. Blackbirds and thrushes sang in the heart of the built-up area, if Wordsworth is to be believed :

" At the corner of Wood Street, when daylight appears,
    Hangs a Thrush that sings loud, it has sung for three years."

The weary pacer along Oxford Street had only to look northwards along the many streets going off into Marylebone to see the open country. In De Quincey's words :

" oftentimes, on moonlight nights, during my first mournful abode in London, my consolation was (if so it could be thought) to gaze from Oxford Street up every avenue in succession which pierces northwards through the heart of Marylebone to the fields and the woods. . . ."

This was the time, too, when a few miles out of London, say on the route followed by John Gilpin, " citizen of credit and renown," and those excellent sportsmen, Piscator, Venator and Auceps, deep country might rapidly be reached, still within sight of the dome of St. Paul's. Clare was not far from the banks of the Lea when he wrote :

> " I love the forest walks and beechen woods
> Where pleasant Stockdale showed me far away
> Wild Enfield Chase and pleasant Edmonton.
> While giant London, known to all the world,
> Was nothing but a guess among the trees,
> Though only half a day from where we stood."

If we have so far rather stressed the debit side of London's spread over the countryside, it is because in this period of great expansion there was little but debit to record. There were, however, already some signs of a backwash movement of animals and plants adapting themselves to city life. In the days when unpaved and uncleansed streets provided an abundance of mud for their nests and an abundance of flies for their food, swallows, house-martins and swifts nested right in the heart of London. Gilbert White, in his famous correspondence with Daines Barrington and Thomas Pennant, mentions house-martins nesting in Fleet Street, the Strand and the Borough, and swifts in the neighbourhood of the Tower. White also noted sand-martins on the outskirts of London, " frequenting the dirty pools in the neighbourhood of St. Georges-Fields and about White-Chapel," and thought they might be nesting in holes in old or deserted buildings. The pigeons of St. Paul's and other parts of London had already adapted themselves to nesting in holes in buildings instead of in the holes of cliffs like their wild progenitors the rock-doves.

THE EFFECTS OF CULTIVATION

During the period between the end of the fifteenth and the beginning of the nineteenth century, the medieval system of agriculture, with its open common fields and its manorial dues and exactions, was completely eliminated in the country round London, as indeed practically everywhere else in the kingdom. The countryside of Middlesex and North Surrey gradually assumed its present familiar patchwork appearance, criss-crossed with hedgerows and so studded with trees, both in the hedges and in the pasture, as to give it the aspect of an all-embracing park. The influence of the hedgerow on the flora and fauna of Southern England has been profound, for both plants and animals, especially birds, have found there a refuge, but for which many of them would have been driven out when the rough grazings of the numerous commons were brought under the plough.

As the City gradually spread over the fields and gardens that lay under its walls, the farming of the remaining land within easy reach became more and more concentrated on serving the special needs of the metropolitan market, and so was comparatively little affected by the turnover to sheep that revolutionised farming in the rest of England from the fourteenth century onwards. The farmland in the immediate neighbourhood of London was increasingly devoted to dairying and market gardening.

As London's population continued to grow, an ever-widening area of the Home Counties was called on to provide food for the Londoners. Norden describes how the countryfolk of Tudor Middlesex earned their livelihood by selling food in the London markets :

" Such as live in the inn countrye, as in the body or hart of the Shire, as also in the borders of the same, for the most part are men of husbandrye, and they wholly dedicate themselves to the manuringe of their lande. And they comonlye are so furnished with kyne that the wyfe or twice or thrice a weeke conveyethe to London mylke, butter, cheese, apples, peares, frumentye, hens, chickens, egges, baken, and a thousand other country drugges which good huswifes can frame and find to get a pennye."

By the beginning of the eighteenth century London's demand for food drew supplies from the whole country. Defoe, setting out on his tour of England in 1724, wrote that :

> " it will be seen how this whole kingdom, as well as the People, as the Land, and even the Sea, in every part of it, are employ'd to furnish something, and I may add, the best of everything, to supply the City of *London* with Provisions . . . Fewel, Timber and Clothes also."

Cobbett hit out with some of his finest invective at the thought of the hated Wen reaching out its tentacles so far :

> " In our way to Swindon, Mr. Tucky's farm exhibited to me what I never saw before, four score oxen, all grazing upon one farm, and all nearly fat ! They were some Devonshire and some Hereford-shire. They were fatting on the grass only, and I should suppose that they are worth, or shortly will be, thirty pounds each. But the great pleasure with which the contemplation of this fine sight was naturally calculated to inspire me was more than counter-balanced by the thought that these fine oxen, this primest of human food, was, aye, every mouthful of it, destined to be devoured in the Wen, and that too, for the far greater part, by the Jews, loan-jobbers, tax-eaters, and their base and prostituted followers . . . literary as well as other wretches, who, if suffered to live at all, ought to partake of nothing but the offal, and ought to come but one cut before the dogs and cats ! " (October 2, 1826.)

Defoe and Cobbett doubtless exaggerated a little, but by the time of the Great Exhibition it was literally true that the food requirements of the London market were drawn from every part of the world then in trade relations with Western Europe.

To return to the influence of London on its own hinterland, Stow, writing in 1598, provides evidence of both the main uses of cultivated land near the City. In a well-known passage he describes a farm at the Minories, near Tower Hill :

> " Near adjoining to this abbey (nunnery of St. Clare), on the south side thereof, was sometime a farm belonging to the said nunnery ; at the which farm I myself in my youth have fetched many a half-

pennyworth of milk, and never had less than three ale pints for a halfpenny in the summer, nor less than one ale quart ɪor a half-penny in the winter, always hot from the kine, as the same was milked and strained. One Trolop, and afterwards Goodman, were the farmers there, and had thirty or forty kine to the pail. Good-man's son being heir to his father's purchase, let out the ground first for grazing of horses, and then for garden-plots, and lived like a gentleman thereby."

A few pages later, Stow mentions the similar fate of a market garden in Houndsditch :

" The residue of the field was for the most part made into a garden by a gardener named Cawsway, one that served the markets with herbs and roots ; and in the last year of King Edward VI the same was parcelled into gardens wherein are now many fair houses of pleasure built."

The fate of Mr. Cawsway's was that of every other market garden that has lain in the path of London's advance. No doubt he, like others after him, retired a mile or two farther out to a fresh holding, unless indeed the proceeds of the sale enabled him, like Mr. Goodman, junior, to live like a gentleman.

By the time John Rocque made his map of the County of Middlesex in 1745, market and nursery gardens covered a large acreage along the Thames between Chelsea and Brentford, while the rest of the county was roughly divisible into two by a line drawn from Edgware to Acton. To the east of this line nearly all the land was enclosed and under grass, with only a few patches of arable on the higher ground. Western Middlesex was largely unenclosed, and still in a very wild state, with extensive stretches of heathland. Hounslow Heath, for instance, then reached almost to Uxbridge. The arable fields in this part of the county, especially around Harrow, were still cultivated on the medieval common field system, though nearer London, around Acton, Chiswick and Hammersmith, the ploughland had been enclosed.

Clearing of waste land and woodland went on steadily through the eighteenth century. Half of Highbury Wood, where Gerard had found the alder buckthorn (*Rhamnus frangula*) in great plenty, was grubbed up as early as 1650, and Cream Hall in Highbury Vale is the farmhouse that was built when the wood was finally cleared about twenty years later. The hills about Hornsey, originally heavily wooded,[1] were cleared at the end of the eighteenth century and laid down to grass ; Highgate Wood and Queen's Wood, formerly known as Gravel Pit Wood and Churchyard Bottom Wood respectively, are all that remain of these woods. When the woods around Hampstead and Highgate, fragments of which still remain in Ken Wood, Bishop's Wood and Turner's Wood, were grubbed up at about the same time, plants such as the tutsan, one of the St. John's worts, were exterminated.

By the end of the eighteenth century there were 73,500 acres of grassland in Middlesex, compared with 20,000 acres of unenclosed and 3000 of enclosed arable. The most important crop was hay, which together with the clover grown as part of the arable rotation was sold in London for horse-feed. A certain amount of corn was also grown, but not much, so that cornfield weeds, such as the corn cockle, have always been relatively scarce in Middlesex. The wheat grown at Heston, however, was reckoned to be the best in the country, and Chelsea, Fulham and Chiswick were famous for their barley. The straw from these crops was also sold into London for the horses.

There were estimated to be over 30,000 horses in London and Middlesex, over a third of them in the cities of London and Westminster, and their dung was carted out to manure the land that provided their fodder. Middlesex-grown fodder also went to feed the dairy-cows of London. At this time milk was produced by cow-keepers rather than dairy farmers, the cows being either kept in byres right in the built-up area or grazed on pastures on the immediate fringe, around Knightsbridge, Edgware Road, Paddington, Gray's Inn Lane, Islington, Hoxton, Mile End, Limehouse, Poplar, Hackney, Bow and Shoreditch. In 1794 there were estimated to be 7200 milch cows in London and Middlesex, and another 1200 in the adjacent parts of Kent and Surrey. The cows were mainly of the short-horned Holderness type, ancestors of our present Shorthorns. Their yield of 730 gallons per year, which compares not unfavourably with present-day yields, was obtained by feeding the cows with turnips, brewers' grains,

---

[1] Hornsey Wood was given by Ray (*Historia Insectorum*, 1710) as a locality for the Duke of Burgundy fritiilary (*Nemeobius lucina*), which was also at that time " pretty common about Dulwich."

best meadow hay, tares and cabbages, as well as the rich grass of their paddocks.

Sheep were unimportant in the rural economy of Middlesex, apart from the luxury trade of rearing lambs by various forcing methods. Many pigs were kept, especially by malt distillers, who fattened them on their grains. Some London citizens retained till quite a late date the ancient right of pasturing pigs and other animals in Epping Forest. Poultry were also kept, especially on the arable farms, where they could pick up much grain that would otherwise have been lost.

Market gardens covered an area of some 10,000 acres in Middlesex at the end of the eighteenth century. London dung was carted out for manure, and came back in the form of spinach, radishes, onions, peas, cabbages, cauliflowers, celery, endive and salads. These market gardens were intensively cultivated with the spade, and provided a rich seed-bed for many weeds, such as the goosefoots (*Chenopodium*) referred to by Curtis, who also remarked of the common meadow grass (*Poa annua*) that it " appears to be the first general covering which nature has provided for a fruitful soil when it has been disturbed." Beyond the belt of intensive cultivation there were some 8000 acres, mostly south of the river, under the plough, and producing vegetables such as peas, turnips and cabbages. At the end of the eighteenth century the market and nursery gardens of Middlesex were mainly found in Chelsea, Brompton, Kensington, Hackney, Dalston, Bow and Mile End, but fifty or sixty years later they had been pushed out to Fulham, Chiswick and Brentford in the west, and Edmonton and Enfield in the north-east.

At one time mulberry trees were grown in London to provide food for silkworms. The first tree planted in England was in the gardens of Syon House, Isleworth, and James I caused many trees to be planted at Hatfield, Theobalds, and on the site of Buckingham Palace gardens. It was stated in *The Field* in 1921 that an old mulberry tree in the grounds of the Palace bore a label, " Planted in 1609 by James I," and it is on record that in 1618 the head-gardener at Theobalds was paid £50 for making a place for the King's silkworms. Later on, Chelsea Park was planted with mulberries for silkworms as a commercial venture, and many of these trees survived till the Park was destroyed to make way for Elm Park Gardens in 1875. Two fine old mulberry trees still flourish in the Chelsea Physic Garden (Plate IX).

At the beginning of the nineteenth century there were still as many as 16,650 acres of common land in Middlesex, but in the next fifty years enclosures proceeded rapidly, following the General Enclosure Act of 1801, until only the few commons that still remain as open spaces were left. The old commons were mostly covered with gorse, heather and bracken, with only about a quarter of their area under grass, and poor grass at that, so that their enclosure had a serious effect on the heathland flora and fauna of Middlesex. At the opening of the nineteenth century such birds as the nightjar, stonechat and Dartford warbler must have been quite common in the county ; by the time Harting wrote in the sixties they were already much reduced in numbers ; to-day the Dartford warbler is extinct as a Middlesex breeding species, and the other two each breed in scarcely half a dozen places in the county.

THE USE OF WILD ANIMALS AND PLANTS FOR FOOD

In Tudor times many wild birds and beasts were still taken for the pot, and if conditions on the Continent even to-day are anything to go by this may well have had a serious effect in keeping down the numbers of small song-birds. An interesting list of the birds consumed at a " principall feeste " in the household of a Tudor nobleman in 1512 has come down to us in the records of the fifth Earl of Northumberland. They included " cranys, redeshankes, fesauntes, sholardes (spoonbills), knottes, bustardes, great byrdes, hearonsewys (herons), bytters (bitterns), reys (reeves), kyrlewes, wegions, dottrells, ternes and smale byrdes." All these birds were probably obtainable within a few miles of the City walls at this time, either in the marshes lining the Thames, or in the case of the bustards probably on wild heaths like Hounslow. Spoonbills have not been known to breed in England since the seventeenth century, but an interesting piece of evidence has come down to us to show that they nested at Fulham in the reign of Henry VIII. In 1523 the Bishop of London sued one of his tenants who had broken into his park and taken herons and " shovelers," as spoonbills were then called, which were nesting in trees there.

Swans were still eaten with relish at banquets in London in the sixteenth century. When the Serjeants of the Inner Temple feasted in 1532, 168 swans were served to their table, and on a later occasion, in 1555, the doughty lawyers polished off 95 swans at a sitting.

Plate XI

Cock black redstart leaving nest-hole in the Temple   ERIC HOSKING

Hen black redstart with food for young   G. K. YEATES

Plate XII

Starlings roosting in plane trees at Marble Arch

The taking of small birds, partly for the pot and partly for sale as cage-birds, continued until prohibited by law in the 1880's. A hundred nightingales are said to have been caught round Harrow by bird-catchers in May, 1838, and an old bird-catcher told Bowdler Sharpe, who wrote in 1894, that he remembered as a boy catching a gross of goldfinches one morning on the site of Paddington Station.

Rabbits, of course, were a perennial source of food. Woolwich Arsenal was built on the site of a rabbit warren in 1720, and there was a regular rabbit warren for London in Epping Forest (Plate 8a) as late as 1821.

Fish also were drawn on abundantly for the table in Tudor and Stuart London. Stow writes of a " great store of very good fish of diuers sortes " in the town ditch under the City wall, but the main source of fish at this time must have been the Thames itself, and an elaborate code of regulations aimed at preventing over-fishing. In the reign of Elizabeth the punishment for fishing offences was made to fit the crime, for Henry Machin recorded in his diary how—

" the xxij day of Marche (1561) dyd a woman ryd a-bowt Chepesyd and Londun for bryngyng yonge frye of divers kynd unlafull, with a garland a-pone her hed with strynges of the small fysse."

As late as the mid-eighteenth century, Robert Binnell, Water Bailiff of the City of London, could

" venture to affirm that there is no river in all Europe that is a better nourisher of its fish, and a more speedy breeder, particularly of the flounder, than is the Thames."

Before the construction of Teddington Lock the flounder, which though a salt-water fish often ascends rivers beyond the tidal limit, was found as far upstream as Hampton Court. Eels used to be taken in large numbers between Hammersmith and Kew, but even in the eighteenth century many were imported from Holland for the London market. A relic of the eel-fishery remains in Eel-Pie Island, an eyot in the river at Twickenham. Southwark plaice remained a delicacy till the late eighteenth century, though latterly they may well have had no more connection with Southwark than Dover soles to-day have with Dover. Even as late as the period 1794-1821 as many as 7346 lb.

of salmon were taken from the Thames, but increasing pollution thereafter drove them out, the last recorded capture being in June, 1833. The shad, once so common as to give its name to Shadwell, also disappeared.

The taking of herbs for medicinal and culinary purposes continued throughout the sixteenth to the eighteenth centuries, and doubtless contributed to the extermination of many species near London. Gerard, as we have seen, recorded how pennyroyal was gathered on Mile End and other commons for sale in London. The great John Ray, in his *Synopsis Britannicarum* in 1690, recorded the sow-thistle (*Sonchus palustris*) growing " near Greenwich and about Blackwall," and added that " some people use this plant when green and tender as a vegetable and in salads ; I leave it to be chewed by rabbits."

Curtis remarked of the dangerous and unwholesome black night-shade (*Solanum nigrum*) that—

> " persons cannot be too nice in selecting their Pot-herbs, particularly those who make a practice of gathering from Dunghills and Gardens a species of Orach, by some called Fat-hen, by others Lambs-quarters, &c., as there is some distant similitude between the two plants, and their places of growth are the same."

This suggests that the gathering of wild herbs for salads and the pot was a regular practice in eighteenth century London. The species of " orach " referred to was probably the white goosefoot or fat hen (*Chenopodium album*), an abundant weed of cultivated ground.

THE WATER SUPPLY

Though the first reservoir for supplying London with water was the great Conduit in Westcheap, begun in 1285, to which led a water-course from Paddington, it was only towards the end of the period under review that the arrangements made for supplying London's water began to have an appreciable biotic effect. When Stow wrote in 1598 there were still many springs and wells in and around London that supplied water to the City, but most of them were already fouled :

> " Besides all which, they had in every street and lane of the city divers fair wells and fresh springs ; and after this manner was this

city then served with sweet and fresh waters, which being since decayed other means have been sought to supply the want. . . ."

One of these means was Lamb's Conduit, which led from the fields behind Gray's Inn, where a street still bears its name, to Snow Hill ; it was constructed in 1577. Thirty-one years later Sir Hugh Middleton began the New River, an aqueduct which still forms part of London's water supply system (Plate 9). The New River has its source at Chadwell spring, half a mile south of Ware in Hertfordshire, but most of the water comes from the River Lea a little below the spring. It is 36 miles long, and ends at the New River Head at the corner of Amwell Street and Pentonville Road, being now bricked in from Canonbury onwards. At one time watercress grew in great profusion under the aqueduct.

The earliest of the Metropolitan Water Board's existing reservoirs was constructed at Stoke Newington in 1834, adjacent to and fed from the New River. Eleven years later the small reservoirs at the north end of Kew Bridge were made. At the end of the eighteenth century a number of important reservoirs had been constructed near London in connection with the new canal system. The Grand Junction Canal, joining the Brent at Hanwell, was begun in 1794, followed in 1801 by the Paddington Canal, which leaves the Grand Junction where it crosses the River Crane, and in 1820 by the Regent's Canal (Plate IV), which links Paddington with the Docks by way of Camden Town. To feed the Grand Junction Canal (now the Grand Union Canal) (Plate V), reservoirs were constructed at Aldenham in 1797, (Plate 10a) and at Ruislip and Kingsbury in 1810.

All these soon became favourite resorts of waterfowl, as to a lesser extent did the canals themselves. The Brent Reservoir at Kingsbury, which was formed by damming the River Brent at the confluence of its two constituents, the Dollis Brook and the Silkstream, is commonly known, from its shape, as the Welsh Harp. In the mid-nineteenth century it became one of the most famous localities in the country for rare waterfowl, thanks to the observations of F. Bond and J. E. Harting. The birds which they and other observers recorded from this one spot up to 1866 included such remarkable rarities as the squacco heron (*Ardeola ralloides*), the night heron (*Nycticorax nycticorax*), the little bittern (*Ixobrychus minutus*), the ferruginous duck (*Aythya nyroca*), the avocet (*Recurvirostra avosetta*), and the grey phalarope, in addition to many other waders, ducks, and gulls at that time recorded from almost no other locality in Middlesex.

In Chapter 12 we shall examine more fully the remarkable effect on the wildfowl population of the London area of having extensive sheets of fresh water, undisturbed by the fowler's gun, within a short distance of the centre of London.

THE GATHERING AND USE OF FUEL

By the opening of the Tudor period London was sufficiently large for its demands for firewood to make considerable inroads on the woodlands of its immediate hinterland. According to Norden the depredations of keepers and commoners so reduced the amount of woodland in Enfield Chase, despite repeated orders from the Crown, that it hardly provided enough fuel for the inhabitants, and we have already noted that cutting firewood for sale in London was a flourishing trade at Enfield at this time. On the other side of London, the Great North Wood, which survives to-day as the suburb of Norwood, was noted for its charcoal-burners. Contemporary writers often jestingly compared the colliers of Croydon with the Prince of Darkness, and it is on record that a collier named (appropriately enough) Grimes, was summoned by Archbishop Grindal in the reign of Edward VI for causing a nuisance with the smoke of his kiln near the present site of Thornton Heath Station. Some two hundred years later a historian of Croydon was still able to write that " the town is surrounded with hills well covered with wood whereof great store of charcoal is made." The association of the district with charcoal-burning is preserved in the name of Colliers Wood, a station on the Underground Railway between Tooting and Morden.

The lopping of hornbeams in Epping Forest and the cutting of furze on Wimbledon Common are two examples of manorial rights which survived right into the nineteenth century because they continued to serve the useful purpose of providing the inhabitants with fuel.

Even at the end of the eighteenth century a great deal of woodland remained in the more rural parts of Middlesex, as well as large untidy hedges, which were allowed to grow to a considerable size and cut back for brushwood every seven or eight years. Marshall, in a survey

PLATE 15

Cabbage attacked by Caterpillars   P. L. EMERY

Red admiral and small white butterflies at Michaelmas daisies   ERIC HOSKING

Comma butterfly on laurel   ERIC HOSKING

with the delightful title of *A Sketch of the Vale of London and an Outline of its Rural Economy*, wrote in 1799 :

> " At present, fagots are of decreasing value in London : owing, I understand, to the bakers having found it cheaper to heat their ovens with coals than with fagot wood, above a certain price : a circumstance which may tend to clear away the pollards and hedge borders, which, at present, disgrace the county."

What saved the woodlands more than anything else was the increasing use of coal for fuel. " Sea-coal " as it was then called, was brought by sea, mainly from the north-east coast, and has left its permanent mark in the name of Sea Coal Lane near Holborn Viaduct, where it used to be unloaded from wharves on the Fleet. Yet while coal reprieved the woodlands till most of them were grubbed up and cultivated, its smoke had a most deleterious effect both on vegetation growing in the City and on human health. John Evelyn complained bitterly of the " hellish and dismall cloud of sea-coale," which had such an effect on the Londoners that " catarrhs, phthisicks, coughs and consumptions rage more in this one City than in the whole Earth besides." Evelyn noted that when the siege of Newcastle in 1644 restricted the supply of coal to London, the gardens flourished unusually well. If he had had his way, all works using coal would have been removed five miles from the City.

By 1718 the menace of the smoke was so great that an official inquiry was held. J. Fairchild, author of *The City Gardener*, published nine years later, stated that certain trees could not be grown in London because of the smoke. Roses were the chief sufferers, whereas holly, ivy, privet and other evergreens were able to stand up to the deposition of soot on their leaves. Even so, as Fairchild's list of plants that could still be grown shows, many more plants flourished in the centre of the early Georgian than in that of the neo-Georgian City two hundred years later :

> " the Lime tree, the Lilac with the white flower—blue and purple, the Persian Jessamin, the Bladder Senna, the Figg Mulberry, Virginia Creeper, Vine, Common Privet, Angelica, Lilies, Perennial Sunflowers, Martagon, John Tradescant's Hare wort, London Pride, Currants, Stock, Guelder Rose, Annuals such as French

Marigold, French Honeysuckle, Pinks, Stocks, Wallflowers, Rockets and Tulips."

Apple-trees grew in the Tower garden, vines outside Temple Bar, and figs in Bridewell and Cripplegate. To-day a vine can still be seen in Gower Street, and fig-trees outside the National Gallery ; Bladder Senna (*Colutea arborescens*) grows in some abundance on a railway embankment at Dalston, and Virginia Creeper can be seen in a good many places in Central London.

It was owing to its habit of shedding its bark at intervals, together with any sooty accretions, that the plane eventually became the typical London tree. The oriental plane (*Platanus orientalis*) was introduced into England in 1562, and the western plane (*P. occidentalis*) in 1636. Our familiar London plane (*P. acerifolia*) is almost certainly a hybrid between the two. Fairchild mentions a plane 40 feet high in the churchyard of St. Dunstan-in-the-East, but the oldest planes now living in London are thought to be those in Berkeley Square, planted in 1789 (Plate 39). In 1906 two of the largest of these were estimated to measure 85 feet high and 13-14 feet across.

THE EFFECTS OF REFUSE DISPOSAL

The Tudor period showed little, if any, improvement on the Middle Ages in the disposal of household refuse and slops. We have seen that many of the springs and streams in and around the City were permanently polluted by matter that nowadays goes down the public sewers, and as London grew the Thames itself became fouled, so that it eventually lost almost all its fish. Laystalls and middens were found in all the main streets, a favourite habitat for many weeds, notably members of the goosefoot family. Gerard noted the stinking goosefoot (*Chenopodium vulvaria*) in some quantity on laystalls, and two hundred years later Curtis recorded dunghills as a habitat of two other goosefoots (*C. album, C. polyspermum*), as well as of the black nightshade.

Throughout the sixteenth century kites seem to have swarmed in London, and being protected as scavengers became so fearless that they would come down and take their carrion even in crowds. Charles Clusius, the great Flemish botanist, who visited England in 1571, thought there were as many kites in London as in Cairo, and described how they picked up and ate the garbage thrown into the streets and even into the Thames. By the eighteenth century they had become rare, having apparently lost their protected status, but in 1734 some still nested in the trees round St. Giles-in-the-Fields, together with rooks and magpies, while Pennant in a letter written in August, 1777, mentions some young kites taken from a nest in Gray's Inn, with frogs in their stomachs. The final disappearance of the kite from London was not chronicled ; from the fact that Pennant also refers to the kite having nested in Hyde Park on two occasions as something rather unusual, it seems probable that kites ceased to breed in London before the end of the century. The story of the kite in London concludes with one seen flying over Piccadilly on June 24, 1859.

Ravens, like kites, lost their privileged status as scavengers some time between the reigns of Elizabeth and George III. Robert Smith, who wrote *The Universal Directory for Destroying Rats and other Kinds of Fourfooted and Winged Vermin* in 1768, relates that he was allowed as much per head for killing ravens as for kites and hawks. Ravens at that time had the reputation, along with the birds of prey, of being killers of young chickens, ducklings and rabbits, and were caught in traps baited with rats. Smith claimed to be able to catch large numbers of ravens in a day, being helped in this by their sociable habit of rallying round when another of the clan was trapped. " London ravens " were easily identified at rabbit warrens in the country by the dirty brown of their plumage, compared with the jet-black of the country ravens, due to their wallowing in the dirt when getting their food. As with the kite there is no record of the latest date at which ravens nested in London, but the last printed record appears to be of a pair that nested in an elm in Hyde Park in 1826, while in Middlesex they are known to have lingered in the Enfield district till the '40's. Later observations of the raven in London, such as one that stayed for some weeks in Kensington Gardens in the spring of 1890, are open to the suspicion of the birds being escapes from captivity.

In spite of the steps taken against them, pigs still scavenged among the laystalls as late as Tudor times. Stow has an anecdote about them :

" And amongst other things observed in my youth, I remember that the officers charged with the oversight of the markets in this city, did divers times take from the market people, pigs starved, or otherwise unwholesome for man's sustenance ; these they slit in the ear. One of the proctor's for St. Anthonie's tied a bell about the neck, and let it feed on the dunghills ; no man would hurt or take them up, but if any gave to them bread, or other feeding, such would they know, watch for, and daily follow, whining till they had somewhat given them ; whereupon was raised a proverb, ' Such an one will follow such an one, and whine as it were an Anthonie pig ; ' but if such a pig grew to be fat, and came to good liking (as oft times they did), then the proctor would take him up to the use of the hospital."

The one beneficial use to which the stinking dunghills of old London were put was the provision of manure for the market gardens on the outskirts of the city. The London dung was first applied to forcing salad crops and raising seedlings in hot-beds, and afterwards, when it had lost its heat and become rotten, dug into the ground in the normal manner.

THE INFLUENCE OF TRADE AND COMMERCE

As the net of commerce was spread ever more widely, various animals and plants arrived in the British Isles as unintended, and sometimes unwelcome, visitors. Many of them stayed to become some of our most tiresome pests of to-day. Rats (Plate VIa), bugs and cockroaches (Plate VIIa) all belong to this category, and as London has always been the greatest port of England, so it was always one of the first disembarkation points for these camp-followers of commerce, as Professor Ritchie calls them.

The brown rat (*Rattus norvegicus*) was the largest, as well as the most destructive, of the undesirable immigrants in this period. A native of the Caspian region, its invasion of Western Europe began on a large scale in 1727, when according to Pallas vast hordes moved westwards after an earthquake, swam the Volga and swarmed across Russia. Thence the brown rat reached England, probably in ships trading with Russia, in 1728 or 1729. To the Englishmen of the day anything unpleasant arriving from the Continent was put down to the new dynasty, so the newcomer was quite inaccurately dubbed the Hanoverian Rat. It did not take the brown rat long to make itself

PLATE 17

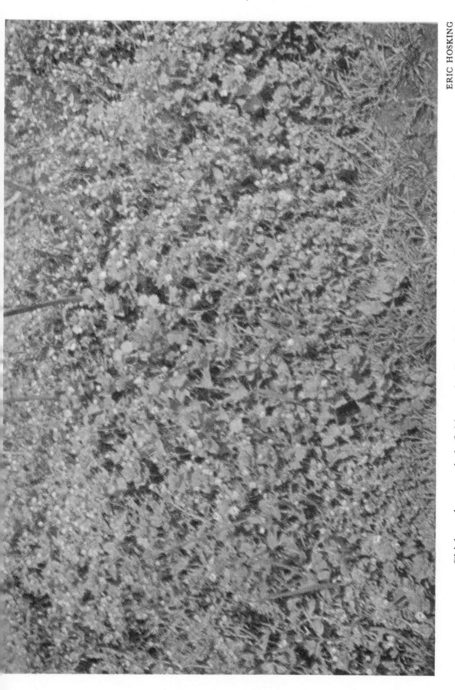

ERIC HOSKING

Chickweed, groundsel, field speedwell, red dead-nettle and annual meadow grass
on market garden land near Chelsfield, Kent

PLATE 18

ERIC HOSKING

The duelling ground at Ken Wood

at home, and as we have seen only forty years after its arrival the trade of ratcatcher was a thriving one, though of course the black rat was already in possession. In fact, the brown rat being stronger, more ferocious and more fecund than its black congener, soon supplanted it over wide areas of the country.

Turning to the invertebrate invaders, cockroaches were first mentioned by Moufet in 1634 as frequenting wine-cellars and flour-mills, and at that time were probably confined to London and a few other ports. This was the common cockroach (*Blatta orientalis*), a native of tropical Asia, which has since been followed by the German, American and Australian cockroaches (*Blatella germanica, Periplaneta americana, P. australasiae*), which are typically found in London hotels, the Zoo and Kew Gardens respectively.

The bed-bug (*Cimex lectularius*), a native of the Levant, was not known in Britain before the sixteenth century, but by 1583 it was causing alarm among ladies of high degree at Mortlake, whose houses were infested with it. While, therefore, bugs may have been imported to London with the wood used for rebuilding after the Great Fire, as tradition insists, it was evidently another case of bringing coals to Newcastle.

That destructive pest of fruit-trees, the so-called American blight or woolly apple aphis (*Eriosoma lanigera*), was first recorded in England in 1787, when Sir Joseph Banks traced it to a nursery in Sloane Street.

An example of an accidental introduction, which so far as we know was not harmful to man or his domestic animals or plants and subsequently died out, was the white-keeled snail (*Helix limbata*), a Mediterranean species, which was reported as common in the hedges round London in 1837.

THE EFFECTS OF SPORT

Sport played an increasingly important part in the relations between man and some of the mammal and bird inhabitants of the London area from the beginning of the Tudor period onwards. Henry VIII took his hunting as seriously as Nero took his fiddling. In 1545, flouting the rights of the citizens of London, he issued a proclamation protecting the game in the district between Westminster and Hampstead :

" Forasmuch as the king's most royall majestie is most desirous to have the games of hare, partridge, pheasant, and heron, preserved in and about his honor att his palace of Westminster, for his owne disport and pastime ; that is to saye, from his said palace of Westminster, to St. Gyles in the Fields ; and from thence to Islington, to our Lady of the Oke ; to Highgate ; to Hornsey Park ; to Hamsted Heath ; and from thence to his said palace of Westminster, to be preserved and kept for his owne disport, pleasure and recreacion ; his highnes therefore straightlie chargeth and comaundeth all and singuler his subjects, of what estate, degree or condicion soever they be, that they, nor any of them, do presume or attempt to hunt, or to hawke, or in any meanes to take, or kill, any of the said games, within the precincts aforesaid, as they tender his favor, and will estchue the imprisonment of their bodies, and further punishment at his majesties will and pleasure."

It was chiefly for hawking that Henry wished to have a large part of the present West End preserved for his own pleasure, and modern Londoners can thank him for ensuring at the same time the preservation of Regent's and other West End royal parks. The king, like his Plantagenet predecessors, kept his falcons at the Mews at Charing Cross, where the National Gallery now stands, till a fire in his stables at Bloomsbury caused him to bring his horses to the Mews. Thereafter the name was transferred from the hawks' home to that of the horses, till in the fullness of time, after the spread of mews all over London as the rich took to carriages, and the later supplanting of horses by motor-cars, it finally came to mean a fashionable habitat for bright young people in the twentieth century.

Royal persons were not the only ones who liked to hunt in the woods and fields of Middlesex. In the reign of Mary Tudor, Henry Lord Berkeley, for instance, " enjoyed daily hunting in Gray's Inn Fields and in all those parts towards Islington and Heygate with his hounds, whereof he hath many and those excellent good ; and there came to share his sport ' many gentlemen of the Innes of Court and others of lower condition ' that daily accompanied him." Stow also recorded that " in hawking and hunting many grave citizens at this present have great delight, and do rather want leisure than goodwill to follow it." At a Citizens' Common Hunt in 1562, for example,—

" the Lord Mayor, Harpur, the Aldermen and divers worshipful
persons, rid to the Conduit-Head before dinner. They hunted the
hare and killed her and thence to dine at the Conduit-Head. The
chamberlain gave them good cheer ; and after dinner they hunted
the fox. There was a great cry for a mile, then the hounds killed
him at St. Giles ; great hallooing at his death and blowing of
horns ; and the Lord Mayor and all his company rode through
London to his place in Lombard Street."

Hares and foxes were minor game, however. The boars and the
deer were the true quarry of kings and lord mayors, but the first of
these were rapidly dying out. The last wild boar in Essex, survivor of
thousands that roamed the marshes and woods in former years, is
said to have been speared by the Earl of Essex at Earl's Colne in the
north of the county in the reign of Elizabeth. South of the river a few
hung on in Windsor Park for James I to hunt them in 1617.

Deer were hardier than boars, and even where they were not
preserved in parks lived on in the woodlands. In Epping Forest,
hunting ground of English kings from Saxon times, the red deer stayed
until 1827, when they were all removed to Windsor. The fallow deer
were reinforced by fresh stock from Norway at the instance of James I,
who was a great hunter ; this suggests that like the roes, which actually
became extinct in the Forest in the sixteenth century, the fallow deer
were in low water.

The following verse celebrating royal hunting in Epping Forest has
come down to us :

> " The second Charles of England,
> Rode forth one Christmastide,
> To hunt a gallant stag of ten,
> Of Chingford Woods the pride."

In aid of this kingly sport the local farmers had to submit to the very
considerable depredations of the deer, which sometimes prevented the
cultivation of the land at all. By the end of the eighteenth century,
however, Epping Forest had become the playground of Londoners it
has remained ever since, and the hunting tradition was reduced to an
annual run with a carted stag.

It is of some interest to recall that Hyde Park was originally emparked and stocked with deer to provide a convenient hunting ground for Henry VIII. James I was graciously pleased to bestow on the Ambassadors of France, Spain, Venice and the Netherlands the privilege of killing a brace of bucks with hounds and bow in this and other royal parks. In 1619 there was a serious poaching affray in Hyde Park, in which two or three poachers were caught shooting the deer at night, and were executed at Hyde Park Gate, together with an unfortunate labourer whom they had hired for 16d. to hold their dogs. To Charles I the Londoners of that time owed the graceful gesture of opening the park to the public for their recreation. The deer, valued at £300, were all sold when the Crown lands were broken up under the Commonwealth, but the stock was replenished at the Restoration, and some stayed on in a paddock till 1831, when they were finally removed on account of the number of pet dogs shot by the keeper for worrying them.

St. James's Park was also stocked with deer by Henry VIII, and these likewise disappeared during the Civil War, though Parliament ordered the park to be restocked from Hampton and Bushy Parks. By 1665 the deer had become part of a menagerie, for Evelyn found there " deere of several countries, white ; spotted like leopards ; antelopes ; an elk ; red deere ; roebucks ; staggs ; Guinea goates ; Arabian shepe, &c." Regent's Park, then known as Marylebone or Marybone Park, was not restocked with deer after the Commonwealth.

The most famous packs of staghounds associated with London were the Royal Buckhounds and the Lord Mayor's staghounds, with which the Citizens' Common Hunt, as described above by Stow, was con- ducted. Latterly they only participated in the carted staghunts in Epping Forest, already mentioned. Carted stags were also hunted by the Berkeley staghounds in the Harrow Weald district. On one occa- sion, as is related in the *Victoria County History of Middlesex*, the stag took refuge in a kitchen, whose indignant owner waved the master's apologies aside with :

" Your stag, sir, not content with walking through every office, has been here, sir, here in my drawing room, sir, whence he proceeded upstairs to the nursery, and damn me, sir, he's now in Mrs. ——'s boudoir."

The Old Berkeley foxhounds were more fortunate in having wild quarry, which on one occasion they are said to have found in Scratch Wood, near Elstree, and lost in rough ground and cover in Kensington. The Berkeley country was reputed to be the most extensive in England, stretching from Scratch Wood and Kensington Gardens all the way to the Berkeley estates in Gloucestershire. The pack seems eventually to have given up hunting in Middlesex, not so much for lack of foxes, but on account of the many people " of lower condition " who tried to attend the meets, as an earlier Berkeley had found. The M.F.H. was obliged to abstain from advertising the meets

" in order to avoid the pressure of a swarm of nondescripts who, starting from every suburb of London, were glad to make a meet of foxhounds their excuse for a holiday on hackney or wagonette, overwhelming the whole procedure by their presence and irritating farmers and landowners, to the great injury of the hunt."

The formal sport of shooting grew in importance during the eighteenth century, and the large bags which were often obtained seriously endangered the stocks of partridges and pheasants. To offset this the red-legged or French partridge was introduced by many land-owners, and we hear of some being bred at Wimbledon prior to 1751, while they are known to have been turned out at Windsor earlier still. Rough shooting remained a popular sport, as is evident from the many records of snipe and woodcock being shot on the outskirts of London mentioned above (p. 65). In a contemporary magazine, in September, 1766, we read that " a gentleman shot a brace of grey Plovers near Mr. Townsend's Menagery in St. George's Fields ; they were esteemed a great rarity at this season of the year and were sent as a present to a noble Duke in Surrey."

A sport much favoured by the " nondescripts " and those " of lower condition " was hunting ducks with spaniels and other dogs—the origin of the well-known inn sign, " The Dog and Duck." The inn of that name in Southwark was a resort of sportsmen who pursued this pastime in the stagnant pools that were to be found in South London till the beginning of the nineteenth century, and the boys of West-minster used to hire dogs to chase the ducks in the marshy pools of Tothill Fields and the Five Fields. Their betters had their duck decoys, one of which was built in St. James's Park by Charles II., who paid a certain Sydrach Hilcus £30 for " contriving " it. The name of Duck

Island commemorates the site to this day. Evelyn noted that "the Parke was at this time stored with numerous flocks of severall sorts of ordinary and extraordinary wild fowle, breeding about the decoy, which for being neere so great a Citty, and among such a concourse of souldiers and people, is a singular and diverting thing."

The Thames remained a good river for anglers till a surprisingly late date. We have seen that a salmon was taken as late as 1833, and a list of the fish frequenting the Thames in Middlesex in 1819 included, in addition to the salmon (*Salmo salar*), trout (*Salmo trutta*), grayling (*Thymallus thymallus*), perch (*Perca fluviatilis*), carp (*Cyprinus carpio*), tench (*Tinca tinca*), roach (*Rutilus rutilus*), dace (*Leuciscus leuciscus*), gudgeon (*Gobio gobio*), pike (*Esox lucius*), eels (*Anguilla anguilla*), and lampreys (*Lampetra fluviatilis*). The Thames being tidal, many salt-water fish were also recorded, including turbot (*Scophthalmus maximus*), sole (*Solea solea*), plaice (*Pleuronectes planessa*), skate (*Raja batis*), halibut (*Hippoglossus hippoglossus*), haddock (*Gadus æglifinus*), oysters, mussels and prawns. Forbes, who wrote the angling section in the *Victoria County History of Middlesex*, tells of an octogenarian relative with memories of the Thames at this time, who assured him that the piles of Old London Bridge were encrusted with mussels, and that the water up to that point, then green and limpid, was quite brackish. Within a generation the water had become so poisoned with sewage that most of these fish had gone from the Inner London reaches of the Thames. Hofland wrote in his *British Angler's Manual* in 1848 that salmon had been driven from the Thames by the gasworks and steam navigation, the last one he had seen being netted at Twickenham in 1818, while trout were few, but "celebrated for their huge size and the excellence of their flavour." From Battersea Bridge upwards, pike and jack were numerous, and perch, barbel (*Barbus barbus*), chub (*Squalius cephalus*), eels, lampreys, flounders (*Platichthys flesus*), roach, dace and gudgeon abundant. Smelts (*Osmerus eperlanus*) could still be taken near London Bridge, as three hundred and fifty odd years before they were taken above Fulham, when Turner wrote to Gesner. The Wet Docks below London Bridge were still mentioned as a good fishing station in 1848.

The Lea was regarded as second only to the Thames by London anglers, and first among these who have fished the Lea must be mentioned Isaac Walton, whose Piscator was on his way to fish at Ware when he stretched his legs up Tottenham Hill to overtake Venator and Auceps one fine fresh May morning. In Hofland's time the Lea above Limehouse ran through " a beautiful pastoral country adorned with villages . . . through parks and meadows containing countless herds of cattle and flocks of sheep," and the fishing between Lea Bridge and Stratford was rented and preserved by a Mr. Beresford of Homerton, who allowed others to fish there for half a guinea a year. According to Hofland, this stretch of the Lea then abounded with pike, jack, perch, chub, barbel, roach, dace, eels, gudgeon and bleak (*Alburnus alburnus*).

AMENITY AND OTHER ÆSTHETIC INFLUENCES

With increasing prosperity in the Tudor period, it was possible to devote a larger proportion of the national income and resources to purely amenity purposes. The most outstanding of these was sport. Hunting and falconry became, even for non-royal personages, activities in which the supply of food resulting from a day's chase was only a secondary consideration, while in fox-hunting and pheasant-shooting the transformation went so far that in the one case a harmful animal was preserved so that it might be hunted, and in the other a voracious bird was artificially bred to provide sport for townsmen.

A similar trend is noticeable in the development of parks and gardens. Parks were originally enclosed to provide a protected grazing place for deer and other hunting quarry, but evolved into purely recreational areas for general public use, while gardens, beginning as places where little else but herbs and vegetables were grown, had already by Tudor times become flower gardens, where both indigenous and exotic plants were grown for purely æsthetic reasons.

St. James's Park, as we have seen, was created out of a marsh for his deer by Henry VIII. Under Charles II it developed into the kind of public park we know to-day. The famous landscape gardener Le Notre planted lime-trees and made a canal, which still survives as the lake. When Evelyn visited the park in 1665, he found many strange waterfowl there, and " withy-potts or nests for the wild fowle to lay their eggs in, a little above the surface of the water." Among the more remarkable birds he saw were a pelican,[1] " a melancholy water-fowl

---

[1] A Pelican (*Pelecanus onocrotalus*), which may have escaped from St. James's Park, was shot at Horsey, East Norfolk, on May 22, 1663, on the testimony of Sir Thomas Browne (" On Norfolk Birds " *Works, 4,* 313-24.)

brought from Astracan by the Russian Ambassador," " a curious sort of poultry not much exceeding the size of a tame pidgeon, with legs so short as their crops seem'd to touch the earth," a white raven, and a crane with a wooden leg which had been made for it by a soldier, and with which it " could walke and use it as well as if it had been natural."

It was not till 1829 that Regent's Park supplanted St. James's as the focus of zoological sightseeing in London. In that year the young Zoological Society of London founded its menagerie, which was much augmented five years later by the addition of the whole remaining collection from the Tower. This had fluctuated much since the Middle Ages, reaching a zenith of eleven lions, two leopards or tigers, two cat-a-mountains or catamounts, a jackal, three eagles and two owls in 1708, and a nadir of one elephant, a grizzly bear and some birds in 1822. Not all the royal animals were kept in the Tower, however, for in addition to the St. James's Park menagerie we know that James I kept ospreys, cormorants and tame otters on the Thames at Westminster.

Hyde Park owes much of its present appearance to Queen Caroline, consort of George II, who in 1733 drained some unwholesome ponds along the course of the river Westbourne, and made the Serpentine and Long Water, where to-day the bathers and boaters and birdwatchers of London all disport themselves. A hundred years later the water of the Westbourne had become so fouled with sewage that it was diverted to a sewer (which to-day can be seen running through Sloane Square Underground Station, above the platforms), and fresh water was supplied to the Serpentine by a water company.

Many exotic trees were introduced into the parks and gardens of London in the seventeenth and eighteenth centuries, in addition to the mulberry, which was planted for mainly utilitarian reasons. The first tamarisk imported into Britain was planted at Fulham Palace by Bishop Grindal in the reign of Elizabeth, and a later bishop, Compton, is said to have been mainly responsible for the popularisation of the American maples, hickories, acacias and magnolias in this country. When Ray visited his gardens at Fulham in September, 1687, Compton was said to have the finest arboretum in England, including cactus, American holly and tulip-tree from Virginia, genista from Africa, and Virginian plane.

Four cedars of Lebanon [1] (*Cedrus libani*) were planted in the

---

[1] Plate VIII shows a fine cedar in Highgate Cemetery.

PLATE 19

ERIC HOSKING

A magnolia in the grounds of Ken Wood House

ERIC HOSKING

The Thames at Hammersmith with mute swans in the foreground

ERIC HOSKING

Teddington Lock

Chelsea Physic Garden about 1683 ; two of them were cut down and sold as timber in 1771 because they kept the sunshine from the flowers, but the remaining two survived till 1878 and 1903 respectively. These trees claim to be the first planted in this country to have firmly established themselves ; they were certainly the first to produce cones, in 1732. The most interesting of all the trees introduced at this time was the maidenhair or ginkgo tree (*Ginkgo biloba*), which was in fact returning to an area it had last inhabited many million years ago. The ginkgo is actually coeval with the dinosaurs in origin, and has only been saved from extinction by its adoption by the Chinese as a sacred tree. It is at present found nowhere in the world in a truly wild state. The famous Kew Gardens ginkgo (Plate 11) was planted about 1760. Others may be seen in Hampstead, Highgate and elsewhere round London.

Many more exotic plants than trees were found in seventeenth and eighteenth century London gardens. To describe these adequately would need a separate book, but we may take Gerard's as an example. Among the many flowers it contained were tulips, wallflowers, Canterbury bells, campanulas, golden rod, snapdragon, sweet william, pyrethrum, periwinkles and marigolds. He made many acclimatisation experiments, both successful and unsuccessful. Sugar-cane, for instance, succumbed to the rigours of a London winter, but he succeeded in growing the crown imperial in " great plentie." Potatoes purchased at the Royal Exchange were a failure, and attempts to grow peaches, apricots and apples suffered because " the poore will breake down our hedge and we shall have the least part of the fruit." Altogether Gerard cultivated over a thousand different species of plants in his garden near the Fleet river.

Many amateur gardeners, having no foreign correspondents to send them exotic plants, went forth into the woods and fields of the London countryside in search of roots, especially of primroses and ferns. The lily of the valley, formerly abundant on Hampstead Heath, was eradicated largely in this way, and by the middle of the nineteenth century even primroses had become quite scarce round London. Not all flower-lovers were so rapacious as a parson named Miles of Cowley near Uxbridge at the end of the eighteenth century, of whom a contemporary said that he " is orchis mad, takes them all up, leaves none to seed, so extirpates all wherever he comes, which is cruel, and deserves chastisement."

A less important kind of depredation on the countryside was the constant demand for evergreens and other leafy boughs, as well as flowers, for all kinds of festive occasions. Clusius observed in March, 1579, that daffodils were sold in great abundance in Cheapside by countrywomen and that all the shops were bright with them. Stow tells how all houses were decorated with holly and ivy at Christmas, and the maypoles all decked with hawthorn boughs on May Day. In the ancient pagan festival of May Day, now transformed into the international workers' day, we can perhaps discern the beginnings of the modern cult of taking an æsthetic pleasure in Nature. According to Stow :

> " In the month of May, namely on Mayday in the morning, every man, except impediment, would walke into the sweete meadows and greene woods, there to rejoice their spirits with the beauty and savour of sweete flowers, and with the harmony of birds, praysing God in their kind."

Other contemporary writers, however, detected something more than a purely philosophical admiration for the beauties of Nature in the maying ceremonies. Philip Stubbes in his *Anatomie of Abuses* complained that :

> " all the young men and maidens, old men and wives, run gadding overnight to the woods, groves, hills and mountains, where they spend all the night in pleasant pastime, and in the morning they return, bringing with them birch and branches of trees and deck their assemblies withall."

Nevertheless, as Oberon's well-known lines show, there was a real appreciation of natural beauty in the early seventeenth century :

> " I know a bank whereon the wild thyme blows,
> Where oxlips and the nodding violet grows
> Quite over-canopied with luscious woodbine,
> With sweet musk-roses and with eglantine : "

At the same time the beginnings of scientific appreciation of Nature were also stirring. From the early seventeenth century onwards the Society of Apothecaries encouraged botanical excursions to see plants growing in their natural habitats as part of the training of apprentice apothecaries. Under the leadership originally of Thomas Johnson rambles were made all over the London countryside, particularly in the Islington-Highgate-Hampstead area, and an account of one of these rambles will be found in Appendix A. In later years the Society's Demonstrator led the parties of apprentices and other students, who met five times a year in the early morning ; greatcoats and umbrellas were banned, but each botanist carried a vasculum over his shoulder, and an attendant followed the party with a larger box for the larger plants. In 1673 the Apothecaries founded their Physic Garden at Chelsea (Plate IX) for the study and cultivation of rare plants and herbs.

The two approaches to the appreciation of Nature, the æsthetic and the scientific, are fused in the celebrated anecdote, which appears to be quite authentic, of Linnaeus falling on his knees in rapture at the sight of the gorse on Putney Heath, when he visited London in 1736. Yet while interest in plants began to be accepted as rational, the unfortunate Lady Glanville, from whom the scarce Glanville fritillary (*Melitæa cinxia*) takes its name, had her will unsuccessfully disputed on the ground that as she had been interested in butterflies she could not have been in her right mind.

We may perhaps fittingly conclude with a pæan by Cobbett in praise of the song-birds of Barn Elms, near Putney, written when he was touring Lincolnshire in 1830, and lamenting the lack of bird song there :

" Oh ! the thousands of linnets all singing together on one tree in the sandhills of Surrey ! Oh ! the carolling in the coppices and the dingles of Hampshire and Sussex and Kent ! At this moment (5 o'clock in the morning) the groves at Barn Elm are echoing with the warbling of thousands upon thousands of birds. The *thrush* begins a little before it is light ; next the *blackbird* ; next the *larks* begin to rise ; all the rest begin the moment the sun gives the signal ; and from the hedges, the bushes, from the middle and the topmost twigs of the trees, comes the singing of endless variety ; from the long dead grass comes the sound of the sweet and soft voice of the *whitethroat* or *nettle-tom*, while the loud and merry song of the *lark* (the songster himself out of sight) seems to descend from the skies."

References to Chapter 5

Ardagh (1928), Bayes (1944), Beadnell (1937), Bell (1924), Besant (1892, 1899), Bishop *et al.* (1928-36), Brett-James (1935), Buxton (1901), Cobbett (1821-32), Cocksedge (1933), Curtis (1777-98), Darby (1936), Defoe (1724), De Quincey (1821), Drewitt (1928), Evelyn (1641-1706), Fairchild (1727), Forbes (1911), Gerard (1597), Glegg (1935, 1939), Graves (1811-21), Gurney (1921), Hamilton (1879), Hampstead Sci. Soc. (1913), Harting (1866, 1880, 1886), Hinton (1931), Hofland (1848), Home (1927), Howard (1943), Johnson (1629, 1632), Laver (1898), Leigh (1821), Loftie (1883), Marshall (1799), Newton (*c.* 1680), Norden (1593), Page (1911), Parkinson (1640), Pena & De Lobel (1570), Perceval (1909), Petiver (1695), Raven (1942), Ritchie (1920), Sharpe (1894-97), Shirley (1867), Smith (1768), Stow (1598), Tallents (1943), Trimen & Dyer (1869), Turner (1548, 1551-66), White (1788), Willatts (1937).

ADDENDUM : Creighton (1891).

THE WEN BURSTS

FROM the middle of the nineteenth century onwards the expansion of London proceeded in almost geometrical progression, as the human population of the metropolitan area grew. By the beginning of the twentieth century the population of the County of London (Map 8) had reached its peak at about 4½ millions, but the out-county area continued to grow apace long after the central core in and around the City (Map 5) had become almost uninhabited at night except for caretakers. By 1931 the population of Greater London, roughly the area within 15 miles of Charing Cross (Maps 10 and 11), had reached over eight millions and showed no sign of slackening its rate of increase. The following table gives the population of London within its various boundaries in 1931, the date of the last published census :

Boundary	Area in Statute Acres	Population	Population per Acre
City of London - - - -	677	10,999	16·2
Administrative County of London -	74,850	4,397,003	58·7
Greater London (Metropolitan and City Police Districts) - - -	443,455	8,203,942	18·5

In the centre of the area occupied by this vast community of human beings, the greatest that has ever been known on the face of the earth, is a solid core where hardly any of the soil ever sees the light of day. Apart from a few pocket-handkerchief churchyards and a handful of exiguous public flower-gardens, the soil, whereon alone a normal plant-growth can be expected, is buried beneath buildings, roads, railway lines and other barren man-made habitats. Few plants except the ubiquitous mosses, and among animals only man's own commensals, such as sparrows, pigeons, rats, mice, spiders, flies, bugs, lice and cock-roaches, are able to find a living in this desert of asphalt and brick.

A glance at the map of the wholly built-up area of London, which is roughly coincident with the agriculturally unproductive land on the

1872

1897

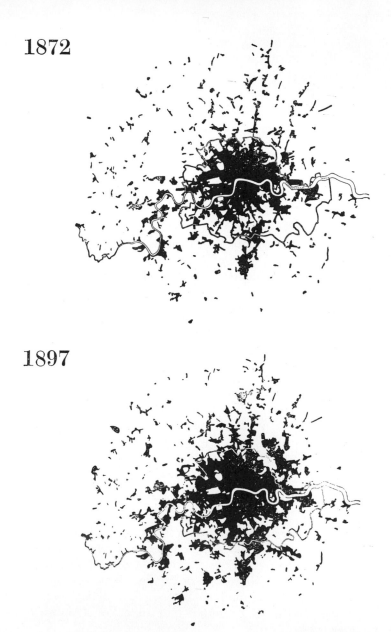

MAP 4 The Spread of London, 1872-1935. The black outlines show the built-up area of London in 1872, 1897, 1914 and 1935. The London

1914

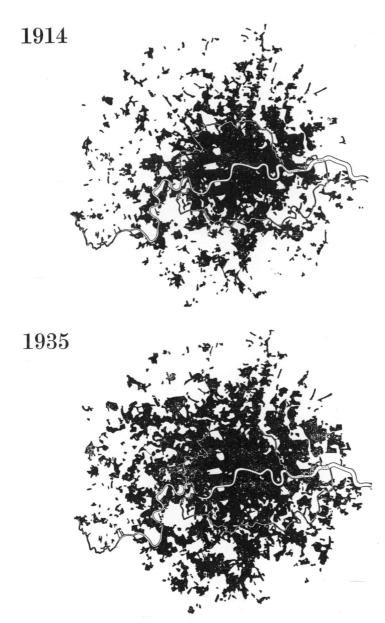

1935

county boundary is also indicated.
(Based on the map of the Growth
of Greater London, published by
the London County Council, 1939)

Land Utilisation Survey Map, shows an irregularly shaped mass (Map 4) clustered on the Thames from Barking and Woolwich in the east to Hammersmith and Putney in the west, a distance of some twelve miles as the crow flies. Northwards there is a stretch of land roughly three miles wide, all along, which is wholly built up except for a few parks and a number of squares and very small gardens, and this sends tongues out for another three or four miles farther still. On the south side of the river the continuous wholly built-up area is less extensive, being confined to a strip two or three miles wide along the river, and is at its thickest within the chord joining Greenwich and Wandsworth. A triangle with a line from Woolwich to Wandsworth as its base, and its apex at Tottenham, contains an area by far the greater part of which is wholly built up.

Outside this inner core is a penumbra, varying in width from five to twelve miles, of partly built-up areas, consisting for the most part of houses with gardens. These two, the wholly and partly built-up areas, constitute a true biological unit within which the influence of man on the natural communities is at its most intense.

The inclusion of the partly-built up areas extends the range of London far beyond the traditional area within which Bow Bells can be heard. Within a radius of twelve or thirteen miles of the City, practically the whole of the land is occupied by roads, railways, buildings, gardens and a relatively small area of parks and other open spaces. Beyond this somewhat jagged circle stretch fingers of houses and gardens pointing along certain main roads and river valleys to the still open country : north up the Lea valley to Cheshunt and Waltham Abbey ; north-east up the Roding valley to Buckhurst Hill and Loughton ; east along the old Roman Colchester road to Romford and Brentwood ; south-east along Watling Street to Crayford and Dartford ; south past Croydon to Caterham and Coulsdon ; south-west to Epsom and Surbiton ; west along the Bath and Oxford roads to Hounslow and Uxbridge ; north-west to Harrow and Edgware and almost to Watford. One by one the villages and market-towns of the Home Counties are being engulfed by the advancing tide, and inexorably linked by bricks and asphalt with Piccadilly Circus and the villages and market-towns on the other side of London.

The counterpart of this process of advance by man has been one of retreat by many of the animals and plants that formerly inhabited the now built-up areas, while others, such as the wood-pigeon and the

Plate XIII

Typical suburban back gardens   JOHN MARKHAM

Plate XIV

Unusual site of blackbird's nest   H. RAIT KERR

Blue tit at coconut   A. R. THOMPSON

rosebay willow-herb, have moved in. An excellent example of the retreating fauna is provided by the history of the rook in London. In the nineteenth century Central London was still near enough to the open fields, where the rooks foraged, for several thriving rookeries to exist on sites completely surrounded by bricks and mortar. In 1836 the largest rookery in London was the one in Kensington Gardens, with about a hundred nests. At this period small colonies and even single pairs would nest wherever there was a suitable tree in the inner built-up part of London. In 1836 a pair nested in Wood Street, Cheapside, and two years later a pair attempted to nest in the crown above the vane of St. Olave's, Hart Street, both in the heart of the City. In 1846 there were fifty nests in the rookery in Chesterfield Gardens, Mayfair, which Harting considered had probably increased at the time of the destruction of the Carlton House Gardens rookery. In 1866 there were rookeries in some trees in a garden at the back of Gower Street and in a large tree in Marylebone Road, opposite Devonshire Place. As late as 1881 there were forty-five nests in the long-established rookery in Holland Park. After this the spread of buildings over their feeding grounds told heavily on the rooks of Inner London. The great rookery in Kensington Gardens was deserted in 1880, largely because the grove of 700 trees in which it was situated was cut down, a senseless act of vandalism that excited the just wrath of W. H. Hudson, who has left us a vivid picture of his first visit to this rookery :

" Never grass and trees in their early spring foliage looked so vividly green, while above the sky was clear and blue as if I had left London leagues behind. As I advanced farther into this wooded space the dull sounds of traffic became fainter, while ahead the continuous noise of many cawing rooks grew louder and louder. I was soon under the rookery listening to and watching the birds as they wrangled with one another, and passed in and out among the trees or soared above their tops. How intensely black they looked amidst the fresh brilliant green of the sunlit foliage ! What wonderfully tall trees were these where the rookery was placed ! . . . Recalling the sensations of delight I experienced then, I can now feel nothing but horror at the thought of the unspeakable barbarity the park authorities were guilty of in destroying this noble grove."

Apart from a single pair that nested in the south-west corner of the Gardens in 1892, and eleven pairs which reappeared for one year in 1893, the rooks repaid this affront from the Commissioners of Woods and Forests by abandoning the site altogether. Up to 1889, however, parties of forty or so could be seen feeding in Hyde Park, whither a few pairs had betaken themselves to the Deputy Ranger's grounds, and from 1891 to 1903 there was a small rookery in Connaught Square, not so very far away in Paddington.

The Gray's Inn rookery was now left as the last in Central London. In 1872 it still held thirty-eight nests, but by 1897 the number had fallen to twenty-four. After 1915 this rookery too was deserted, due, it was said, to the birds being disturbed by the drilling of recruits in the Inn, though the major cause was certainly the distance the rooks now had to travel to reach open country to feed. The Gray's Inn Benchers once nearly lost their rooks by an unconsidered act almost as crass as that of the Kensington Gardens authorities. Somebody, evidently ignorant of the right time to lop trees as well as of the psychology of rooks, ordered the large horizontal branches of the trees in the rookery to be lopped off in March to improve their appearance. March being the rooks' busiest breeding month, they were seized with a panic, and departed *en masse*, not returning for three years.

The last recorded occasion when rooks nested in Inner London was in 1916, when four pairs, probably wanderers from the deserted Gray's Inn rookery, raised the hopes of the Templars by recolonising for that one year one of London's most ancient rookery sites. Both Goldsmith and Lamb wrote about the Temple rookery, which is known to have existed as far back as the Great Fire, but disappeared before 1831.

At the present day there are very few rookeries within ten miles of Charing Cross, the nearest on the south side being two small ones at Lee Green, $5\frac{1}{2}$ miles south-east (Plate 12), and Ham, nine miles south-west, while in Middlesex none have been reported recently nearer than Bushy Park, Southall and Mill Hill, all between nine and twelve miles away.

While the disappearance of the rook was only to be expected as a result of the spread of London, it is sad also to have to record the virtual loss of that most urban and ecclesiastical of birds, the jackdaw. Apart from a handful of pairs in the south-west corner of Kensington Gardens, which still drag out a somewhat flaccid existence—nearly fifty years ago Hudson described them as " far less loquacious and more sedate in manner than daws are wont to be "—the jackdaw must rank as a lost London bird.

A hundred years ago daws were not at all uncommon in the centre of London. They frequented the towers of churches, notably St. Michael's, Cornhill, and the Grosvenor Chapel, South Audley Street. Even fifty years ago a dozen pairs nested regularly in the grounds of Devonshire House, Piccadilly, but they were not seen after 1893. They hung on longer in the suburbs, and there also favoured churches as nesting sites. On one occasion, when alterations were being made at the Baptist Chapel in Heath Street, Hampstead, two or three cartloads of sticks, the relics of many jackdaws' nests, were found in the towers.

As the jackdaw is to-day a common breeder in the more rural parts of the London area, nesting in holes in trees and quarries as well as in churches and other buildings, and is not infrequently seen flying over the central area, we can only conclude that it is the difficulty of finding enough food within easy reach of Central London that has driven it, like the rook, away from an area which still abounds in suitable nesting sites.

Many of the smaller passerine and other terrestrial birds that a hundred years ago could be found nesting within half an hour's walk of the West End have also failed to adapt themselves. Whereas some birds, such as the blackbird, song-thrush, robin and blue tit have stayed on in the parks and larger gardens, others like the nuthatch, grass-hopper-warbler, reed-warbler, Dartford warbler, nightingale and cuckoo must now be sought some way out. Many of them, however, still pay fleeting visits to the parks in the central area on migration.

Yarrell recorded nuthatches breeding in Kensington Gardens in 1843, but a quarter of a century later they had become only casual visitors, and to-day they are very rare indeed, the most recent record being of one that visited a garden in Holland Park Avenue on January 17, 1937. Since nuthatches still nest regularly in Ken Wood, which is little over four miles from both St. Paul's and Charing Cross, it seems a little odd that birds so inoffensive and unpersecuted should so early have deserted Inner London. Ken Wood to-day can hardly be regarded as a lonely spot, so that love of solitude cannot be the reason. The explanation is more likely to be found in the sooting over of the trees in Inner London, and this may also account for the disappearance of other birds for which both nesting sites and a certain amount of food are apparently available in the Central Parks. The nuthatch is essentially a bird of woodland and parkland, so that every time an estate with many trees is broken up by the builder, the bird suffers a contraction of range.

The grasshopper-warbler, known chiefly by its penetrating " song," more like an angler's reel than its insect namesake, used once to frequent the Five Fields, where Belgravia now stands, and in the 60's was tolerably common about Hampstead and Highgate. To-day you could not reckon on finding it nearer than Ruislip or Epsom Commons, both over fifteen miles from Charing Cross, and this is undoubtedly due to the extensive draining and building over of the damp heathy places it favours.

The reed-warbler, which is even more attached to water in the breeding season, has also been pushed out by the growth of London. At one time, along with the long-departed bearded tit (*Panurus biarmicus*), it must have nested in the reeds right along the banks of the Thames into London, and even to-day it can be found just outside the county boundary at Chiswick Eyot. Around London, as elsewhere in Britain, it is especially associated with the common reed (*Phragmites communis*), and small colonies may be found almost wherever there is a sufficient growth of this reed. The reed-warbler often breeds at the flooded gravel-pits that stud the gravelly parts of the London area, at Barn Elms and Beddington in Surrey, for instance, and Sewardstone in the Lea valley. Occasionally reed-warblers nest in gardens a long distance from water, and then are especially fond of lilac bushes. In 1864 four pairs of reed-warblers nested in gardens in Hampstead, and the same phenomenon has been observed at Ealing, Teddington and Hampton Court.

The Dartford warbler, though it is not known ever to have come nearer Central London than Hampstead Heath, is another bird that has had its breeding haunts much reduced by the onslaught of the builder. First discovered as a British breeding bird by Latham at Dartford and Bexley in Kent in 1773, this our only resident warbler was found comparatively commonly at Hampstead, Highgate and Finchley in the 60's, and had also been shot on Old Oak Common and Wormwood Scrubs. To-day the Dartford warbler has long vanished from its ·Kentish and Middlesex haunts, and though within the past ten years it has been seen as near London as Richmond Park and Wimbledon Common and has nested in two or three localities in Surrey within the twenty-mile radius, it is not to be found breeding regularly nearer than the heathlands of North-west Surrey, and even there it is only hanging on by the skin of its teeth. The Dartford warbler is very vulnerable to many human activities not directly aimed at it, and in addition suffers severely from hard weather. Building development, egg-collecting, gorse fires and military manœuvres have all contributed to its (possibly temporary) extermination in the London area, but the actual *coup de grâce* was probably administered by the series of severe winters from 1938-39 to 1941-42.

It is many years now since a real nightingale sang in Berkeley Square, though it is on record that in 1703, when the Duke of Buckingham leased the site of the palace that now bears his name, he found there " a little wilderness full of Blackbirds and Nightingales." In 1830 nightingales were said to be not uncommon between Hyde Park Corner and Kensington gravel pits, and in the 60's they could be found in such suburbs as Stamford Hill, Stoke Newington and Victoria Park. To-day the nightingale is rarely heard within a dozen miles of Charing Cross. The most recent record for Inner London seems to be one that was watched in full song in Kensington Gardens by Mr. Holte Macpherson on April 25, 1936, and the nearest breeding place to Central London is probably Ham Common, Surrey. Farther out, in such places as Bookham and Effingham Commons in Surrey, Bricket Wood Common, Herts, and Epping Forest, nightingales still abound.

Several stories are told of insomnia due to the incessant singing of nightingales near London. One victim was Edward the Confessor at his hunting lodge at Havering-atte-Bower, Essex, whence he succeeded in inducing the Deity, so legend relates, to banish the offending birds. A Victorian M.P. was less successful in his prayers, and had to abandon his tenancy of a lodge in Enfield Chase on the ground that his family could get no sleep for the singing of the nightingales.

The furry caterpillars of London gardens have reason to be thankful that cuckoos no longer frequent the inner areas, though in most years they pass through the parks on migration. As recently as 1905 a young cuckoo was reared in a robin's nest in a garden in Marylebone Road, and four years later a cock robin was seen feeding a young cuckoo in the grounds of the Zoo in Regent's Park. The explanation of the disappearance of the cuckoo from Inner London is probably to be found in the absence of its two favourite fosterers, the meadow-pipit and reed-warbler, and the scarcity of the two second favourites, the pied wagtail and hedge-sparrow, leaving only the robin as a possible fosterer. It is significant that where cuckoos have laid eggs in Inner London in the past fifty years they have usually chosen robins as fosterers, and that a young cuckoo reared in a Hampstead garden in 1940 had such unusual foster parents as blackbirds. It would probably be hard for a female cuckoo to find anywhere in Inner London enough robins' and hedge-sparrows' nests sufficiently near together to enable her to lay her normal clutch of a dozen eggs. Since the scarcity of fosterers is directly due to building development, the cuckoo must also go down on the list of birds driven from London by bricks and mortar.

In spite of the steady advance of the built-up areas, however, a remarkable number of different species of bird still contrive to breed in the quite extensive areas that remain relatively open within twenty miles of London. In Appendix B will be found a list of the ninety-nine species of birds which regularly breed in the London area (seventy-three as residents and twenty-six as summer visitors) and of the eighty-eight additional species which visit the area more or less regularly as winter visitors or passage migrants.

The story of the influence of the spread of London on the flowering plants is much the same as that of its influence on the avifauna. Dr. E. J. Salisbury has listed some fifty-five to sixty plants which have either become extinct or have very seriously decreased in Middlesex as a result mainly of the building over and draining of the county. (See Appendix C.) On the other hand there are still several hundred plants (the exact number depends on whether you are a splitter or a lumper in respect of such genera as the roses, brambles and hawkweeds) to be found within twenty miles of London, and Appendix D lists one hundred plants that are to be found in all the twenty-four divisions into which the London Natural History Society has divided its area, and so may claim to be the hundred commonest London plants.

References for Chapter 6

Bishop *et al.* (1928-36), Fitter & Parrinder (1944), Glegg (1935), Harting (1866), Hudson (1898), London Natural History Society (1900-44, 1937-44), Morgan (1939), Salisbury (1927), Yarrell (1843).

NATURE INDOORS

IN ANY human community there are a number of animals and plants that have succeeded in adapting themselves to living actually in or on either the human beings or their dwellings. With the parasitic fauna and flora that lives inside the human body—viruses, bacteria and tape-worms, for instance—and produce diseases, we are not concerned here. Suffice it to say that advances in sanitary science have eliminated a great many of the sources of the spread of bacteria, popularly known as " germs," which in former times led to the great epidemics of plague, cholera and typhoid. The pasteurising of milk has almost banished the scourge of infantile diarrhœa, which even forty years ago carried off hundreds of babies every year in London alone, owing to the prevalence of flies. The careful inspection of all meat coming into London has eliminated the danger of people eating tape-worms in their Sunday dinner.

The parasites that live outside the human body are a different matter. The great evacuation in September, 1939, revealed to a shocked country that head-lice (*Pediculus capitis*) were a great deal commoner than most people had liked to think. Body-lice (*P. vestimenti*), it is true, have almost disappeared in London, but in 1935, for instance, 14 per cent of the elementary school children of the County of London were found to be infested with live head-lice or their eggs. It is well known that certain families living in slum or semi-slum conditions form permanent reservoirs of lice, regularly reinfesting the children who have been cleansed at school. Fleas (*Pulex irritans*) are nowadays a relatively minor problem, though they are still occasionally to be picked up in public places, such as a cinema or bus. These are as likely to be dog-fleas as human ones, owing to the frequency with which people take their dogs about with them.

Whereas lice and fleas inhabit the human body and clothes all the time, bugs only feed there, retiring to cracks in the walls and furniture to rest. There is still a great deal of bug-infested property in London, and in one Kennington slum, for instance, cleared not long ago 98 per cent of the removals had to be disinfested. As slum clearance progresses the old bug-infested property is being gradually destroyed, but unfortunately, owing to inefficient disinfestation, or careless or deliberate avoidance of disinfestation, bugs are often transferred to the new houses when former slum tenants move.

Turning to the more numerous class of commensals which inhabit man's dwellings rather than man himself, we find creatures as large as rats and mice. Rats are usually found only in old and rather dilapidated houses, especially near rubbish-dumps, sewer outlets and warehouses, which are the most favoured habitats of London rats, and where they live in colonies of considerable size. It is related that some years ago, when a large number of rat-infested houses were demolished in London, the hungry rats swarmed into a nearby restaurant, where they competed with the diners for food to the extent of even jumping on to the tables and snatching morsels off the plates. Mr. Hinton also describes how in another London restaurant, not far away, he saw three rats emerge from a chimney and leap on to the floor across a large and fiercely glowing fire. Nearly all the rats found in houses are brown rats (Plate VIa), which prefer cellars and drains to the walls, roofs and ceilings of warehouses which are the habitat of the arboreal black rat.

Not the least important reason for keeping rats out of our homes is the danger of their carrying infection. Rats are well known as carriers of bubonic plague, which is spread by their fleas ; if one of the fleas happens to bite a man instead of a rat, one of the rare cases of plague that have occurred even in the present century in the Port of London may result. The dangerous worm that causes trichinosis in man (*Trichina spiralis*) is primarily a rat-dweller, but if a man eats the imperfectly cooked flesh of a pig that has eaten a trichinosed rat, he also is liable to get the worm. A single infected pig that slipped through the inspection once caused 337 cases of human trichinosis, of which 101 died. The dwarf tape-worm (*Hymenolepis nana*), a common human parasite, is also primarily rat-borne, and this by no means terminates the case against the rat as a carrier of disease.

Mice are much more widely distributed than rats, and few houses in London can boast of being entirely free from house-mice scampering along the wainscoting. In the more rural suburbs various kinds of

Plate XV

A grey squirrel in a London park   JOHN MARKHAM

House-sparrows feeding from the hand in a London park

Plate XVI

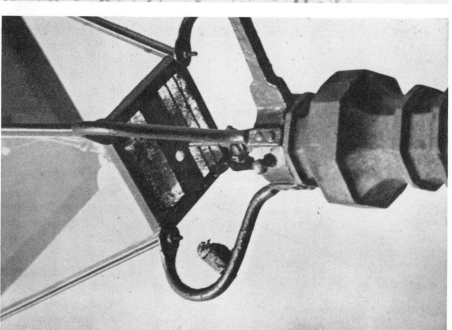

Woodpigeon on nest in plane tree near the Bank of England

Blue tit carrying food to nest in lamp-post   JOHN MARKHAM

field-mice will occasionally enter houses, but in the centre all the mice found in houses are the true house-mouse (*Mus musculus*) (Plate VIb).

The only other mammals which actually live indoors are bats. Mr. Johnson relates how, when St. Paul's Cathedral was being filmed in connection with the Restoration Fund about 1926, large numbers of bats were revealed by the beams of the flash-lights. Disturbed by the unaccustomed brightness, they fluttered about the cathedral, seeking new roosting places. Elsewhere in Central London, it is thought that bats roost in holes in the old trees in the parks rather than in buildings.

A wide and unpleasant variety of arthropods form the remainder of the indoor animal community. All of them are in greater or less degree unwelcome to their human hosts. They include cockroaches, crickets, ants, spiders, flies, bluebottles and clothes-moths. Some, such as the spiders, eat others, such as the flies, but muscivorous habits do not seem to the average housewife to be an adequate compensation for spiders' webs.

The common cockroach (*Blatta orientalis*) and the German cockroach (*Blatella germanica*) are the kinds most likely to be found in ordinary dwelling houses (Plate VIIa). It is hardly necessary to point out here how undesirable cockroaches are in the house, how quickly they manage to get around to new property, especially to new blocks of flats with furnaces for central heating, or how foolish it is to leave any food uncovered where there are cockroaches. For some unexplained reason, cockroaches, which are brown orthoptera, are popularly known as " black beetles "—beetles being coleoptera.

House-crickets (*Gryllulus domesticus*) may be expected in warm places indoors, especially in crannies near the furnaces of bakehouses and central heating plants. They are particularly common in some of the warmer houses of the London Zoo. Being aliens from Africa, though established in Britain for many centuries, they cannot live out of doors for any length of time, except in the hottest weather and in places, such as rubbish-heaps, where artificial heat is generated.

Several species of ant are found in or near houses, and in London the small red house-ant (*Monomerium pharaonis*) is a common inhabitant of eating-houses ; a tiny introduced ant from the Argentine is found in similar situations.

The arachnologist finds a rich fauna in the houses of London (Plate VIIb). In dwelling houses there is the common house-spider

(*Tegenaria derhamii*), with its unpleasantly large relatives, *T. atrica* and *T. parietina*, the latter of which is known as the Cardinal, from a possibly apocryphal occasion when it frightened the great Wolsey. Spiders with more specialised habitats include *Leptyphantes nebulosus*, which likes warehouses, and the introduced *Theridion tepidiarorum*, which prefers hothouses and warm greenhouses.

Among the most undesirable of all the hangers-on from the insect world are the several species of flies, which often come to our houses straight from refuse-heaps and dustbins, and alight on our food, some just to lick, others to lay their eggs, which in due course result in fly-blown meat or fish. The chief danger, apart from the general unsavouriness of the proceedings, is that they carry disease bacteria, such as that of infantile diarrhœa already mentioned. The common house-fly (*Musca domestica*) and the bluebottles or blow-flies (*Calliphora erythrocephala, C. vomitoria*) are among the most frequent offenders.

A study of the literature relating to clothes-moths reveals a quite alarming number of species of these microlepidoptera that regularly inhabit our homes and attack our clothes. Most of them are accidental introductions, distributed by commercial intercourse all over the world. The common clothes-moth (*Tineola bisselliella*) is said to have originated in Africa, and the large pale clothes-moth (*Tinea pallescentella*) probably in South America. The common clothes-moth is probably the most destructive of them all, if destructiveness is measured by the amount of damage it does to clothes, furs, carpets, curtains and upholstery each year ; its omnivorous larvæ will eat anything from pemmican to cobwebs. The white-tip clothes moth (*Trichophaga tapetzella*) used to be particularly fond of harness-rooms and neglected coach-houses, where its larvæ regaled themselves on cushions and horse-blankets. It does not seem to find garages so favourable a habitat. This moth derives its name from its remarkably close resemblance to a small bird's droppings, which has led entomologists to suppose that it was not originally an indoor species, but one that has adapted itself to domestic life without losing a characteristic that obviously has more protective value in the open.

The white-shouldered house-moth (*Endrosis lactella*) has larvæ that eat a wide variety of vegetable substances, but not cloth, e.g., dry seeds, corks, rubbish in birds' nests, straw thatch, fungi on trees, meal and almost any dry vegetable refuse. It has even been known to eat into the corks of bottled wines, causing them to leak. Another particularly undesirable moth to have about the house is the all too common brown house-moth (*Borkenhausia pseudospretella*), which has been stigmatised as the greatest general pest of all the clothes and house-moths. It has been known to do considerable damage to the bindings of books. Fortunately, clothes and house-moths, like most other insects, have their own private pests, and several species of ichneumon fly have been proved to parasitise them.

Among other insects that either inhabit or invade houses in London may be mentioned the silver-fish (*Lepisma*), a primitive wingless insect that hides in cracks, disused cupboards and chests of drawers. It is sometimes accompanied by the beetle *Niptus hololeucus*. Other beetles among the domestic fauna include *Anobium panicum* in bread and groceries generally, and various furniture beetles, including the famous death-watch beetle (*A. domesticum*), which is especially harmful to old woodwork, such as that of Westminster Hall, and of which Keats wrote :

> " Make not your rosary of yew-berries,
> Nor let the beetle, nor the death-moth be
> Your mournful Psyche, . . ."

Compared with the fauna of domestic interiors, the indoor flora is happily very meagre. Various moulds and fungi attack food that is left too long in a warm, damp place, dry rot (e.g. *Merulius lachrymans*) destroys old woodwork, and in very damp and slummy conditions moulds and fungi may also appear on walls and furniture, but for all practical purposes we may leave the plant kingdom till we get out of doors.

References for Chapter 7

Austen & Hughes (1932), Hinton (1931), Johnson (1930), Ministry of Agriculture & Fisheries (1937), Ritchie (1920), Sinclair (1937).

THE WHOLLY BUILT-UP AREAS

FROM a biological point of view perhaps the most important effect of the spread of London has been to create what is probably the largest area in the world of an entirely new type of habitat, the wholly built-up area. Over many acres of Central London the ground is completely covered with buildings, roads and railway lines, and vegetation is almost non-existent (Plates II, Xa ; Maps 5, 6). Yet even in this desert, quite a number of animals, especially birds, have managed to adapt themselves to the novel environment.

The only mammals that frequent the wholly built-up areas, apart from man himself, and his domestics, cats, dogs and horses, are the rats and mice, those ubiquitous hangers-on of human communities. They also attract their own enemies, for besides the cats, owls will stray into the city streets at night in search of a meal. No other mammals, and no amphibians or reptiles, have succeeded in adapting themselves to the peculiar conditions of life in Inner London, except where parks, gardens or squares offer some alleviation of the barrenness.

Birds are the largest creatures which have succeeded in fully adapting themselves to outdoor life in the centres of great cities ; their superior mobility enables them to escape from the many dangers of town life by flying up to the roof-tops. Even so, only two species are really typical of the wholly built-up areas of London, the house-sparrow and the semi-feral London pigeon, both of which can support themselves without ever setting eyes on a blade of grass or a green leaf. Admittedly there are many parts of the central built-up areas where an odd tree or two has managed to survive, but these occasional trees are by no means essential to the existence of the sparrows and pigeons.

A few other birds are at any rate partial inhabitants of the wholly built-up areas. Gulls come to forage for what they can get, but they always depart in the evening to roost on one of the great reservoirs in

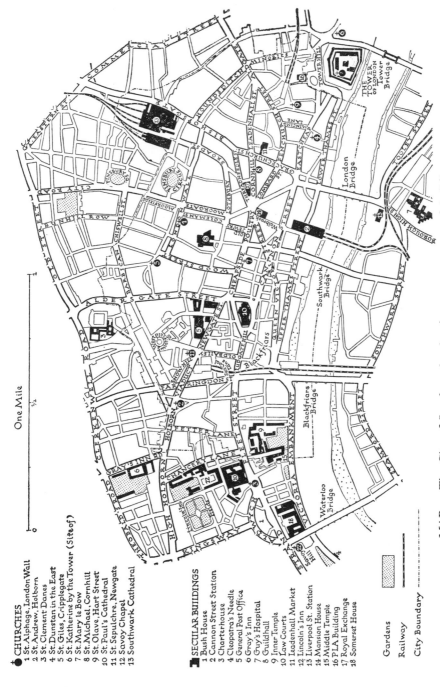

One Mile

MAP 5 The City of London, showing important streets and buildings. (Based on the 6" Ordnance Survey Map, Edition 1920) *By permission of H.M. Stationery Office*

the Thames or Lea valleys. Kestrels nest on such buildings as West-minster Abbey and the Imperial Institute in South Kensington, but their foraging areas include the parks and other partly built-up areas. The peregrine falcons that occasionally visit St. Paul's and the Palace of Westminster—the most recent record is of one perching on Big Ben on May 27, 1942—are more truly urban birds, as they must prey largely on the pigeons. Black redstarts have recently colonised Central London, but though they nest on buildings they are always dependent on some kind of waste ground on which to feed. Swifts have a good claim to be included on the list, for they feed in the air, and in recent years have been seen wheeling over the East End and Bloomsbury, Marylebone and Westminster in the summer. No evidence, however, has been produced that they breed in the central built-up area ; Barnes, Notting Hill, Kilburn and Hampstead prob-ably represent the inner limit of their breeding range to the west and north-west of London. Carrion-crows are never seen in the inner areas, except in the parks or " as the crow flies " from one part of London to another. Starlings, on the other hand, are birds which have within living memory adapted themselves to roosting on buildings in London, though returning each morning to the suburbs to feed.

The house-sparrow is the London bird *par excellence*. It is the only bird allowed by general consent to be a Cockney, and certainly the Londoner's favourite, as witness W. H. Hudson's anecdote of the men who came to feed the sparrows in the Dell in Hyde Park :

> " ' I call these my chickens, and I'm obliged to come every day to feed them,' said a paralytic-looking, white-haired old man in the shabbiest clothes, one evening as I stood there ; then, taking some fragments of stale bread from his pockets, he began feeding the sparrows, and while doing so he chuckled with delight, and looked round from time to time, to see if the others were enjoying the spectacle.
>
> " To him succeeded two sedate-looking labourers, big, strong men, with tired, dusty faces, on their way home from work. Each produced from his coat-pocket a little store of fragments of bread and meat, saved from the midday meal, carefully wrapped up in a piece of newspaper. After bestowing their scraps on the little brown-coated crowd, one spoke : ' Come on, mate, they've had it all, and now let's go home and see what the missus has got for *our*

tea ' ; and home they trudged across the park, with hearts refreshed and lightened, no doubt, to be succeeded by others and still others, London workmen and their wives and children, until the sun had set and the birds were all gone."

Naturally, London sparrows do not depend wholly on human largesse for their sustenance, though kind-hearted people must feed many loaves of stale and not so stale bread to them every day. In the central built-up areas they forage all over the streets and buildings, railway sidings and waste plots, wherever there is any kind of animal or vegetable food to be picked up, for the house-sparrow is omnivorous. *The Handbook of British Birds* gives its diet as " in towns, insects, street refuse, and few seeds." The sparrow population of Central London must have been seriously reduced by the decrease in the amount of horse-drawn traffic in the past forty years. Not only were the horses a source of food with their droppings and the sweepings from their nosebags, but the fast-moving and petrol-fumed motor traffic does not allow the sparrows much time to forage in the roadways for the food scattered by the few remaining horses.

Far too little attention has been paid to the habits of the London sparrows, and in particular to their local roosts or " chapels," which were noted as far back as 1865. Some are in trees, like the one outside the British Museum, but many are on buildings. In 1867 some 200 sparrows were knocked down and drowned by a severe rainstorm in one roost in a garden in Bethnal Green. The extreme localness of most of the roosts is shown by the fact that hardly any definite fly-lines have been traced towards them.

Pigeons (Plate 7) were noted in London as early as 1385, and they are now abundant residents, being even more truly birds of the built-up areas than sparrows, for while sparrows are also found in the partly built-up areas, pigeons are almost exclusively denizens of the wholly built-up areas. They treat the towering buildings of modern London as cliffs, thereby giving an additional indication of their descent from the wild rock-doves (*Columba livia*) that still frequent our north-western coasts. Only exceptionally do London pigeons perch in trees ; normally they rest either on the ground or on the ledges of buildings. They nest in small colonies in all sorts of holes and crannies in buildings, usually high up and inaccessible, just as their wild rock relatives will seek the remotest crevices for their nesting places.

The London pigeon has been most unfairly and cavalierly treated by most writers on the birds of London. Although it is one of the most numerous species within the bounds of the metropolis, its claim to full status as a resident London bird have been either ignored or square-bracketed, just because (like the little owl, mute swan, Canada goose, pheasant and French partridge) it happens to be descended from birds escaped from captivity. While it may well be that the stock is constantly being reinforced by fresh escapes from pigeon-lofts round London, there can be no doubt at all that for more than five centuries pigeons have been living in London in as nearly a feral condition as is possible for any bird living in a great city. It is argued that the pigeons are regularly fed by the public in such places as Trafalgar Square and in front of St. Paul's, but there are thousands of pigeons in the back streets that have to scrounge for themselves, and in any case we do not deny wild status to the birds that come to the bird tables or to the gulls that are fed on the Embankment. A more serious objection is that the London pigeon has not yet established its type, being still a hotchpotch of various domestic strains. This is true, though the amount of white in the London pigeons seems to be decreasing, and the whole population is gradually reverting to something like the original rock-dove plumage. It cannot be maintained, however, that this is an objection to recognising the London pigeon as a wild bird, since it is in fact living as a wild bird ; it is only an objection to giving it a scientific name and a separate specific or subspecific status.

It is not often that London acquires a new breeding bird, and still less often that the newcomer elects to breed in the heart of the built-up area. Yet this is what has happened since 1940, when a pair of black redstarts brought off two broods in the precincts of Westminster Abbey. The story of the black redstart (Plates XIa, XIb) in the London area, which is quite the most remarkable in the annals of London ornithology in the past half-century, actually begins some years earlier. It has only quite recently come to light that black redstarts began to breed in the Palace of Engineering at Wembley in 1926, the year after the end of the Empire Exhibition. At this time the bird was known only as a scarce winter visitor and passage migrant in the London area, and it was only three years since it had first been found nesting in the British Isles, on the Sussex coast.

We now know that three pairs of black redstarts nested in the Palace of Engineering in every year from 1926 to 1941, a fourth pair

ERIC HOSKING

A black-headed gull being fed in St. James's Park

W. SUSCHITZKY

Waterfowl in St. James's Park — South African grey-headed sheld-duck,
a pair of mallard and a coot

PLATE 22

ERIC HOSKING

Sheep grazing in front of Ken Wood House

also nesting in 1937. Here year after year, all unknown to the numerous ornithologists who live in and around London, in an almost deserted factory building, but close to one of Britain's biggest annual concourses of people—at the Wembley Stadium on Cup Final day— these rarest of British breeding birds reared their two broods of young. The nests were always built on similar breeze-slab ledges about 18 feet above roller doors that are large enough to admit a lorry, but are protected by lofty porticoes. The birds ceased to breed in the Palace of Engineering in 1942, when for the first time, as a result of intensive war work, the roller doors beneath the ledges were in constant use. They have not, however, been driven right away, as cock birds were heard singing nearby in 1942-43, and in 1944 one pair nested in the old site, which was soon afterwards destroyed by a flying bomb.

Meanwhile the black redstart was slowly spreading as a breeding species in other parts of the country. In 1927 it was first seen in Cornwall, and nesting was proved two years later. In 1927 also the bird paid its first summertime visit to Inner London, when a cock and hen were seen, though never together, at the Natural History Museum in South Kensington on four different dates between April and September. It now seems at least possible that they were scions of the Wembley stock. In 1930 occurred the first Kentish breeding record, followed three years later by the first breeding record for the County of London. In 1933 a pair of black redstarts nested in Woolwich Arsenal, rearing their brood in the corner of the eaves of a small shed about 10 feet from the ground in one of the busiest parts of the Arsenal, where all kinds of traffic passed all day long.

In 1936 a period of more intensive colonisation began. In that year a pair first bred in Cambridge, which became one of the chief strongholds of the species, and two separate non-breeding birds appeared in Inner London, where they have occurred in every subsequent year, except for 1937. The 1936 birds both appeared on the doorsteps, as it were, of ornithologists who were able to identify them from previous experience, so that one is left wondering how many more were present, even as early as 1936, that went unnoticed. A cock was seen at the river's edge near Lambeth Bridge on April 29, and subsequently was often heard by Mr. E. M. Nicholson, singing from tall office buildings near his home in Marsham Street, Westminster, till mid-June. The second bird, a hen, was captured in the botany department of the Natural History Museum, South Kensington, on May 19, and subsequently released.

The next summer record of a black redstart in London was in St. John's Wood, where a cock stayed for two and a half hours in a garden on June 22, 1938. In the following year two birds were seen on the back of the British Museum and on the new London University building opposite in Bloomsbury. It is noteworthy that there was a cleared site in the near vicinity, just as there had been in the case of the Westminster bird in 1936. There can be little doubt that these birds fed on the waste ground forming the sites of the London University and Westminster Hospital buildings respectively. Within eighteen months enemy action was to provide a superabundance of such waste sites in Central London.

In 1940 some half-dozen non-breeding cocks were singing in various parts of London, in addition to the pair that bred at Westminster, rearing its second brood in a nest on a ledge behind the old school gate of Westminster School. The other birds were seen in places as far apart as St. Paul's Cathedral, Trafalgar Square, Mayfair, South Kensington and the White City, Shepherd's Bush. Some of these birds may have been identical, however, as the unmated males have a habit of shifting their song-posts and territories fairly frequently. It is important to notice that this large increase took place several months before the " blitz " devastated London.

In 1941 two pairs of black redstarts are known to have nested in London, but only three other singing males were recorded. One pair returned to Westminster, rearing its first brood in the cowl of a stove-pipe 50 feet above the ground and its second in the same nest that was used for the second brood in 1940. The other pair bred in the fireplace of a bombed building in Wandsworth. The non-breeding birds were recorded from Westminster, South Kensington and St. Paul's. A young male at Hillingdon in May and an old male at Hayes in September, both in Middlesex, may have come from Wembley.

The *annus mirabilis* of the black redstarts in London was 1942. A score of singing males were present and at least three pairs bred successfully. At the same time a considerable spread of the bird's range took place over the rest of Southern England. Though breeding outside London was only proved at Cambridge, Whitstable, Kent, and Burlescombe, East Devon, non-breeding birds appeared in places as far apart as Plymouth, Maidstone, Sheffield, Lowestoft, Southampton,

Ely and St. Leonards-on-Sea. In London the only nest actually located was in Wandsworth, where Brigadier Christie saw a nest in the same site as in 1941. Col. Meinertzhagen saw a pair with three young feeding on ants on waste ground in Notting Hill during August, and Mr. H. G. Gould saw the hen of a pair that had frequented Cripplegate churchyard in the City throughout the summer accompanied by three young birds on September 7. The great majority of the non-breeding birds were in the City, where the extensive areas of waste ground and ruins were undoubtedly the primary factor in their ecology. There were four singing males in the area around Fenchurch Street and Mincing Lane, two near Cannon Street Station, and two or three in the area between the Law Courts, the Temple and Fetter Lane. In the West End one returned to the Abbey area, but did not stay, and others were seen in such old and new haunts as St. James's and Hanover Squares, Tottenham Court Road, Baker Street, Earl's Court, Campden Hill and South Kensington. In Middlesex males were seen or heard at Hayes and Southall without evidence of breeding being obtained.

In 1943 the black redstarts maintained their position well, and showed a tendency to spread to the suburbs. Three pairs nested, in the Charterhouse, in the Temple and near Fetter Lane, the two latter both bringing off two broods. More than a dozen other singing males were present in Inner London, and others appeared in the suburbs as far out as Wembley and Ravenscourt Park in the west and Plaistow, Surrey Docks, Isle of Dogs and Purfleet in the east. The non-breeding males of Inner London comprised about eight in the heavily blitzed City area, two or three in Westminster, two in Bloomsbury and one at Earl's Court. For the first time since 1939 no black redstart appeared at the Natural History Museum, South Kensington.

The Charterhouse brood of three was ringed on June 11, and one of these was picked up dead at New Southgate, Middlesex, seven or eight miles away to the north-north-west on July 26. Had it not been ringed, this would doubtless have been adduced as evidence that black redstarts had also bred in or near New Southgate.

In 1944 there were again three pairs known to have bred in Inner London, one in the Temple (Plate XIa), one in the ruined area behind Guildhall, and one in Whitechapel, but considerably fewer non-breeding males were recorded.

It is sometimes said that the " blitz " was responsible for the

A   Albert Memorial, Kensington Gardens
N   Nelson Column, Trafalgar Square
H.G.P. Horse Guards' Parade
O.P.Y. Old Palace Yard, Westminster
S.G. Spring Gardens, S.W. 1

CHURCHES

1 St. Giles in the Fields
2 Grosvenor Chapel, S. Audley Street
3 St. Martins in the Fields
4 St. Marylebone
5 Westminster Abbey
6 Westminster Cathedral

SECULAR BUILDINGS

7   Bethlehem Hospital (Bedlam)
8   British Museum
9   Buckingham Palace
10 Charing Cross Station
11 Chelsea Hospital
12 Devonshire House, Piccadilly
13 Euston Station
14 Fulham Palace
15 Great Central Hotel
16 Houses of Parliament
17 Imperial Institute
18 King's Cross Station
19 Marylebone Station
20 National Gallery
21 Natural History Museum
22 Royal Opera House
23 Paddington Station
24 St. Pancras Station
25 Temple Lodge, Kensington Gds.
26 University of London
27 Victoria Station
28 Waterloo Station
29 Westminster City School
30 Westminster Hospital
31 York Water Gate

MAP 6   The West End, showing important streets and buildings. (Based on the 3″ Map of London,

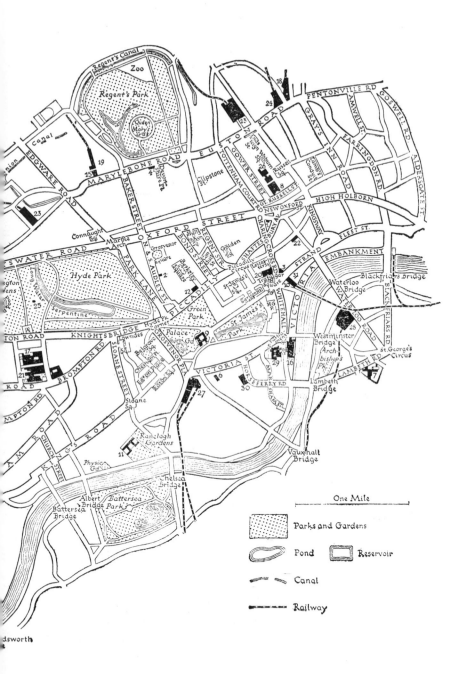

Regent's Canal

Zoo

Regent's Park

Queen Mary's Gdn

Regent's Park

Canal

19

EDGWARE ROAD

25

MARYLEBONE ROAD

BAKER STREET

Devonshire St.

4

PENTONVILLE RD

AMWELL ST.

GOSWELL RD

GRAY'S INN RD

FARRINGDON RD.

ALDERSGATE ST.

EUSTON ROAD

TOTTENHAM COURT RD.

GOWER STREET

High St.

Woburn Sq.

Russell Sq.

23

24

18

GT. RUSSELL ST.

8

NEW OXFORD

HIGH HOLBORN

KINGSWAY

FLEET ST.

Clipstone St.

Grosvenor Sq.

OXFORD STREET

REGENT ST.

BOND STREET

S. & N. AUDLEY ST.

Golden Sq.

Hanover Sq.

PICCADILLY

HATTON GDN.

CHARING CROSS RD.

22

STRAND

EMBANKMENT

Blackfriars Bridge

Connaught Sq.

Marble Arch

Hyde Pk. Corner

Grosvenor Gdns.

Berkeley Sq.

Piccadilly Circus

Leicester Sq.

Trafalgar Square

Waterloo Bridge

BLACKFRIARS RD.

BAYSWATER ROAD

Long Water

Hyde Park

PARK LANE

Serpentine

HYDE PARK

GREEN PARK

ST. JAMES'S

VICTORIA

WHITEHALL

St. James's Park

WATERLOO RD.

25

A

Palace

Gd.

9

St. George's Circus

KNIGHTSBRIDGE

KNIGHTSBRIDGE

Lowndes Sq.

SLOANE STREET

Belgrave Sq.

Westminster Bridge

Arch bishop's Pk.

7

LAMBETH RD.

Lambeth Bridge

17

21

BROMPTON RD.

CHESHAM PLACE

EATON SQ.

VICTORIA ST.

12

13

20

10

28

29

16

5

6

HORSEFERRY RD.

REGENCY ST.

30

27

Sloane Sq.

KING'S ROAD

BROMPTON RD.

CHURCH STREET

Ranelagh Gardens

Physic Gdn.

11

Vauxhall Bridge

Chelsea Bridge

Albert Bridge

Battersea Park

Battersea Bridge

dsworth

One Mile

Parks and Gardens

Pond

Reservoir

Canal

Railway

published by the Ordnance Survey,
1933) *By permission of H.M.
Stationery Office*

125

breeding of the black redstart in London. This is clearly not true, however, for the process of colonisation had begun long before the war, and was in fact part of a general spread which has carried the species over most of north-west Europe in the past 150 years. The most that can be said for the " blitz " theory is that the bombing provided many more nesting sites than would otherwise have been available, and also ensured an abundance of the waste spaces which the bird requires to provide it with the necessary insect food. The war may also have made identification easier, for there is greater audibility due to less traffic, and greater visibility due to the destruction of buildings. If black redstarts could nest at Wembley for so long unsuspected, it is more than likely that they did so elsewhere, and there is every reason to suppose that additional records for past years will continue to turn up from time to time.

One of the most noticeable changes in bird habits in Inner London in recent years has been the development by starlings of the habit of roosting on trees and buildings during the autumn and winter months (Plate XII.) London is not the only British city where this has happened ; starlings roost in the centre of Manchester, Liverpool, Edinburgh, Glasgow, Belfast, and many other large towns. But in London the phenomenon occurs on a much larger scale than elsewhere, just because London is a much larger city, and the fact that the starlings roost on many famous buildings, such as St. Paul's Cathedral and the National Gallery, gives them a news value on a national plane.

It used to be thought that this great influx of starlings into London must be due to immigrants from the Continent, and we are much indebted to the researches of Mr. E. M. Nicholson for exploding this plausible thesis. The starlings that roost in London come, in fact, from no farther afield than the suburbs, where they gather into small groups towards sundown and fly into the centre, gradually forming larger and larger flocks, till on arriving at their roosts they create the illusion that they are great flocks that have flown *en masse* all the way from the outer suburbs. Mr. Nicholson has described what really happens :

" It was particularly interesting to observe the behaviour of starlings in Kensington Gardens towards roosting-time. In the late afternoon they mounted to the crowns of the trees and sang unceasingly to themselves or to each other. One by one they became impatient, and took wing towards the Long Water, where on the fringe of trees they used to halt to form loquacious gatherings in the upper branches. The babble of their conversation attracted many passing birds to descend and join them. By this process its volume rapidly swelled, casting an irresistible spell upon all the starlings within hearing. Nothing so fascinates a starling as the sound of other starlings ; the gathering flock exercised over a spreading radius a kind of magnetism or centripetal attraction, till all the solo performers had come into it. The largest of these assemblies was in the trees between the Peter Pan statue and Temple Lodge. When it had attained a respectable size, parties began to break off from it and fly over the bridge and down the Serpentine, till the departures exceeded the fresh arrivals, and the main flock, taking wing in a body, ascended towards Hyde Park Corner."

The starlings which roost in Central London, which Mr. Nicholson estimated in 1933 to be not more than 20,000 in all, come from suburban areas all round the City, up to a distance of 12-13 miles away. In some suburban areas the starlings fly away from the City to roost at points as near the centre as eight miles, and it is only along the Thames and Lea valleys that they fly inwards from as far out as a dozen miles. During the day the starlings that roost in London feed on all kinds of open spaces, parks, gardens, playing fields, and especially sewage farms.

Aldous Huxley has given a good description in *Antic Hay* of the arrival of the starlings at a small roost in a Paddington square :

" On fine evenings he [Mr. Gumbril] used to sit out on his balcony waiting for the coming of the birds. And just at sunset, when the sky was most golden, there would be a twittering overhead, and the black, innumerable flocks of starlings would come sweeping across on their way from their daily haunts to their roosting-places, chosen so capriciously among the tree-planted squares and gardens of the city and so tenaciously retained, year after year, to the exclusion of every other place. Why his fourteen plane trees should have been chosen, Mr. Gumbril could never imagine. There were plenty of larger and more umbrageous gardens all round ; but they remained birdless, while every evening, from the larger flocks, a faithful legion detached itself to settle clamorously among his trees. They sat and chattered till the sun went down and twilight was past, with intervals every now and then of silence that fell suddenly and inexplicably on all the birds at once, lasted through a few seconds of thrilling suspense, to end as suddenly and sense-lessly in an outburst of the same loud and simultaneous conversation."

The roosts used by starlings in London at present can be classified in the following five main types :

(1) Autumn roosts, mostly on islands in parks outside the central built-up area, where starlings began the habit of roosting in London, and which are now used only, if at all, as gathering places for birds proceeding farther into the centre ; e.g., the islands in the Serpentine and the Regent's Park lake.

(2) Autumn roosts in trees in squares and elsewhere, which are mostly forsaken for building-roosts when the leaves fall ; e.g., Leicester Square and Duck Island in St. James's Park.

(3) Roosts on buildings, usually on ledges or on the acanthus leaves of Corinthian capitals, which are used throughout the roosting season ; e.g., the Royal Exchange, the National Gallery and St. Martin in the Fields church.

(4) Small roosts in such places as church steeples, mostly away from the central area and probably used only by local birds ; e.g., Albert Bridge and Westminster City School.

(5) Summer roosts used by non-breeding birds during the breeding season, the only examples so far recorded being Duck Island and some of the buildings round Trafalgar Square.

The great majority of the roosts are of types (2) and (3), that is to say tree-roosts and building-roosts, and are situated in two relatively small areas in Central London. One of these areas is centred on Charing Cross, and extends north to the junction of Shaftesbury Avenue and High Holborn, east to Somerset House, south to Parliament Square and west to Piccadilly Circus. The other lies in the heart of the City, and extends north to Finsbury Circus, east to the Tower, south across the river to Guy's Hospital and west to St. Paul's Cathedral and St. Sepulchre's, Newgate Street. A secondary area lies along Marylebone Road between the Great Central Hotel and St. Marylebone parish

ERIC HOSKING

Coldharbour Farm, Mottingham — the last farm in London

ERIC HOSKING

A plum orchard near Chelsfield, Kent

W. BENNETT

Goldfinch on nest

ERIC HOSKING

A pair of mute swans at their nest by the River Lea
near Hertingfordbury, Herts.

church, extending south to Marble Arch. (Plate XII.) Since there is a constant interchange of birds between the two main roosting areas, birds that fly in from the east passing over the City to the Charing Cross area and *vice versa,* practically the whole of the north bank of the river between Westminster and the Tower can be regarded as one gigantic starling roost. As a result nearly all the famous buildings of London have at once time or another been used as starling roosts, including the British Museum, the Royal Opera House in Covent Garden, the General Post Office, Guildhall, the Marble Arch, St. Mary-le-Bow, St. Pancras and Charing Cross Stations and the Port of London Authority Building on Tower Hill. Tree-roosts have also occurred in famous places ; there was a very large one, deserted since 1934, in the Savoy Churchyard when the B.B.C. was at Savoy Hill.

One of the remarkable features of the London starling roost system is its relative stability since Mr. Nicholson first mapped it in 1925-26, only a few years after its establishment. This survey revealed that there were three zones in " Starling-London," an outer zone, where they either stayed all the time or flew away from the City to roost, surrounding an inner zone, where they fed by day but flew into the centre to roost at night, and a central core where starlings roosted in large numbers at night but very few spent the day. The actual boundaries of these zones probably fluctuate from year to year, but the general structure of the system remains stable. The only important change noted between 1925-26 and 1932-33, for instance, was the abandonment of the roost on the British Museum.

It is interesting to trace the growth of this new habit of the London starlings. Until the early 90's starlings were known as common breeding birds in the London parks, but the majority left town at the end of the summer and wintered in the country. The first mention of them roosting in London was in a letter to *The Times* of November 3, 1894, from T. D. Pigott, who pointed out that starlings were coming into London every evening to roost in the trees of Duck Island in St. James's Park. When Hudson wrote four years later, starlings were roosting on the islands in the lakes in Regent's Park, Battersea Park, Hyde Park and the grounds of Buckingham Palace, as well as on Duck Island, and were apparently increasing in numbers. At this time most of them still left their London roosts at the beginning of October, and spent the winter at roosts in the country. It seems possible that it was the engulfing of these country roosts by the spread of London that led the starlings to seek roosts inside instead of outside London.

Exactly when starlings began to roost in large numbers on buildings in London is not known, but the habit seems to have been developed gradually in the early years of the present century. The steps taken to drive the birds away when they became too numerous in St. James's Park may possibly have been a contributory factor. When starlings first began to roost on the Nelson Column, they came in ones and twos, instead of in flocks as at present, which seems to confirm the gradualness. Eric Parker gives hearsay evidence that the British Museum and St. Paul's began to be used as roosts during the 1914-18 war, and he is himself the first reliable observer who actually saw starlings roosting on any building in London—on the British Museum in November, 1919, and on St. Paul's in the following January. By 1922 starlings were roosting all over the central area, wherever there were a few trees. The Temple and the Savoy Churchyard were said to be the most crowded roosts, and the National Gallery and Nelson Column were also in use.

Thus some time between 1898 and 1919, the starlings roosting in trees on the islands in the parks, and gradually spreading to the trees in the squares, like Mr. Gumbril's, which is based on Huxley's personal experiences in Westbourne Square before 1914, lost the habit of migrating to the country roosts when the leaves fell, and adapted themselves to roosting on the buildings in the neighbourhood of their urban tree-roosts. It is possible that this transition first took place from Duck Island to the buildings surrounding Trafalgar Square, for at this point an island-roost is nearer a building-roost than at any other. It should be borne in mind that in parts of its range, such as the north of Scotland, the starling is pre-eminently a cliff-dweller, and we have already seen how readily the cliff-loving pigeons have adapted themselves to London. It is worth noting also that to a large extent the tree-roosts in London are still forsaken every autumn when the leaves fall, the birds simply migrating to a building-roost nearby. This is particularly noticeable in Trafalgar Square, where the birds gradually leave the trees round the square as the leaves get fewer and fewer, and join their fellows on the National Gallery and other adjacent buildings.

Like other animals, starlings are most conservative. Each bird has its own niche on the roosting ledge, and James Fisher has shown how, when misfortune overtakes one bird, days or weeks may elapse before the gap in the roosting line is filled.

Apart from the rats and mice and birds, the outdoor fauna of the wholly built-up areas is not particularly numerous or noteworthy. Such as there are are mainly insects, various kinds of flies that flourish on refuse, horse-droppings and similar unsavoury paraphenomena of town-life, ants, crickets, mosquitoes, and a few beetles, such as *Blaps*, the cellar or graveyard beetle. Sometimes the ants force themselves on the attention of Londoners by all swarming on the same day, as on July 23, 1917, when according to Mr. Johnson, pavements as far apart as the Temple Station and Fulham, Westbourne Park and Clapham, were bespattered with crushed ants trodden down by unsuspecting pedestrians. Luckily for Londoners bee-swarming incidents, such as the following related by Pigott, are rare :

" On a Saturday morning in July, 1885, an assistant in Messrs. Mappin and Webb's shop, while crossing the pavement in Regent's Street, found himself suddenly covered from head to waist by a swarm of bees. Fortunately he had presence of mind and kept still until, with the help of sympathetic bystanders, coat and hat were taken off, when, as suddenly as it had come, the swarm rose and left him with no hurt more serious than a couple of stings on the neck."

The cabbage butterflies (*Pieris brassicæ, P. rapæ*) (Plate 16a) that can often be seen fluttering about the streets of London may either be immigrants from the Continent, or have hatched out from a consignment of greenstuffs at the vegetable markets ; in a few cases, no doubt, they have come from cruciferous plants growing in one of the many small gardens tucked away in the back streets. The small brown moths seen fluttering about the streets in spring are usually male vapourers (*Orgyia antiqua*), whose caterpillars, with grotesque tufts of hair, are especially fond of the plane-trees, which line many London streets. The female moths of this species are wingless. Whence came the giant dragon-fly (*Æschna grandis*), measuring four inches in wing-span, that was once seen from an office in Queen Victoria Street, it is impossible to say.

The flora of the wholly built-up areas is even more exiguous than the fauna, which is not surprising since we have defined the wholly built-up areas as those where there is no bare soil for flora to grow in. Nevertheless, even without the intervention of the Luftwaffe, buildings were pulled down from time to time in the built-up areas, and for a space there was opportunity for plants to grow. The London rocket (*Sisymbrium irio*) sprang up in great profusion among the ruins of London after the Great Fire of 1666. Its place as the dominant plant of waste ground has since been taken by the rose-bay willow-herb (*Epilobium angustifolium*), which sprang up abundantly on the site of Bush House after the clearing of the Wych Street area in connection with the Aldwych-Kingsway scheme, and now has become the typical plant of the blitzed areas. A most interesting waste site is the one between Great Russell Street and New Oxford Street, near the Y.W.C.A. building, which has stood vacant for some years and now holds a flourishing young cherry-tree, as well as a patch of bracken (*Pteridium aquilinum*).

London botanists have shamefully neglected the opportunities for studying the flora of the waste sites that lie under their noses, and up to the war the most recent study published was one by Shenstone of five sites in Farringdon Street, Bloomsbury, the City, Covent Garden and Clerkenwell in 1912. The five sites had been cleared between 1901 and 1910, and yielded seventy-eight species of plants, which Shenstone divided into four classes, those distributed by the wind, those with small seeds probably distributed by birds, those probably originating from horses' nosebags, and those originating as garden escapes or from kitchen refuse.

The plants whose seeds were wind-borne included three willow-herbs (*Epilobium angustifolium, E. montanum, E. roseum*), the coltsfoot (*Tussilago farfara*), dandelion (*Taraxacum officinale*), groundsel (*Senecio vulgaris*), two ragworts (*S. viscosus, S. jacobæa*), the common creeping thistle (*Cirsium arvense*), two sow-thistles (*Sonchus arvensis, S. oleraceus*), the introduced Canadian fleabane (*Erigeron canadensis*) and the black poplar (*Populus nigra*). These were well represented on each site. The plants with bird-borne seeds were fewer, comprising five common weeds, shepherd's purse (*Capsella bursa-pastoris*), swine's-cress or wart-cress (*Senebiera coronopus*), mouse-ear chickweed (*Cerastium vulgatum*), common chickweed (*Stellaria media*) and knotgrass (*Polygonum aviculare*). The forage or packing-case plants, considered to have been distributed mainly through the medium of horses' nose-bags, were a much more numerous class, including some thirty-four species, ten of them grasses.

PLATE 25

Anglers on the River Lea near Broxbourne, Herts.   ERIC HOSKING

A bold red deer at Richmond Park   C. E. MANEY

Feeding the pelicans at St. James's Park   ERIC HOSKING

Among them were many common weeds, such as the stinging nettle (*Urtica dioica*), three plantains (*Plantago*), two docks (*Rumex*), two polygonums, the common orache (*Atriplex patula*), and the white goose-foot or fat-hen (*Chenopodium album*), with some typical fodder plants, such as the melilots (*Melilotus officinalis, M. alba*) and Dutch clover (*Trifolium repens*), and cornfield weeds, like the white campion (*Lychnis alba*) and penny-cress (*Thlaspi arvense*). The fourth class consisted of twenty-four miscellaneous plants, most of them escapes from various kinds of cultivation, some from flower-gardens, such as the opium poppy (*Papaver somniferum*) and the evening primrose (*Oenothera biennis*), others from vegetable gardens, like the rape (*Brassica napus*), cabbage (*B. oleracea*) and parsley (*Carum petroselinum*), or from various commercial uses, like *Carthamus tinctorius*, an Indian dye used for adulterating saffron. Such plants as the strawberries and fig-saplings may perhaps be attributed to office-workers' lunches, but one wonders how the deadly nightshade (*Atropa belladonna*) and wood-sorrel (*Oxalis acetosella*) came to be there. It is of some interest that the London rocket was found on the site in Bloomsbury, near the British Museum ; this appears to be the last published record of its occurrence in London.

The numerous wall flora of Gerard's London has almost all gone, though as has been mentioned a luxuriant colony of pellitory (*Parietaria officinalis*) still grows on the piece of London wall in the street of that name near St. Alphage's church. The only other plants likely to be found on other walls in the City are such mosses as *Tortula muralis*, and perhaps an occasional small fern growing in some damp grating.

References for Chapter 8

Calvert *et al.* (1944), Fitter (1943, 1943*a*, 1944, 1944*a*, 1944*b*), Glegg (1935), Hudson (1898), Johnson (1930), London Natural History Society (1937-44), Nicholson (1926), Parker (1941), Pigott (1902), Shenstone (1912), Witherby & Fitter (1942), Witherby *et al.* (1938-41).

THE PARTLY BUILT-UP AREAS:
GARDENS AND PARKS

" A cuckoo called at Lincoln's Inn
Last April ; in Soho was heard
The missel-thrush with throat of glee,
And nightingales at Battersea ! "
LAURENCE HOUSMAN.

AS YOU proceed out from the centre of London, and the value of land
falls, small gardens begin to appear (Plates Xb, XIII, 31a), until in
the outer suburbs many houses have half an acre or more of garden
attached to them. This is the area coloured purple on the Land
Utilisation Map, and forms a transitional zone between the wholly
built-up areas, coloured pink, and the open country, coloured green,
yellow and brown, which is a biologically quite distinct habitat. Interspersed with the houses and gardens are a number of open spaces, allotments (Plate 13a), parks and commons, which provide sanctuary for
certain species for which gardens are too confined in space. The partly
built-up areas have developed a distinctive wild flora and fauna,
which is quite apart from the dogs, cats, chickens, flowers, vegetables
and shrubs that man has brought in to make up his own domesticated
community.

Except for the ubiquitous rats and mice, no mammals have fully
adapted themselves to garden life, though in the outer suburbs field
mice (usually *Apodemus sylvaticus*) are sometimes found. Hedgehogs
(*Erinaceus europæus*) also occasionally wander in from the fields ; Lord
Rosebery saw one in the garden of Mr. Henry Chaplin's house in Park
Lane in 1895, and in the present war they go to drink in flooded
Anderson shelters, and have occasionally been found drowned in one.
There are no reptiles in the garden association (fortunately, most people
will no doubt wish to add), and the only amphibians are an occasional
frog (*Rana temporaria*) or toad (*Bufo bufo*), particularly the latter.

As with the fully built-up areas, it is the birds which have most fully succeeded in adapting themselves to the garden environment. Even such essentially man-shy species as the heron and the kingfisher will visit a small garden in the early morning if there is a pond with goldfish. These, however, are but passers-by. The true birds of the garden association, that can be found in most gardens of any size in the London area, are the starling, greenfinch, chaffinch, house-sparrow, great and blue tits, mistle and song-thrushes, blackbird (Plate 14, XIVa), robin, hedge-sparrow, wren, house-martin and swift. Most of these birds are the ones that householders hope to attract by putting up a bird-table or hanging up coconuts (Plate XIVb) or lumps of fat. A great many other species, such as the linnet, bullfinch, pied wagtail, spotted flycatcher, willow-warbler, great spotted wood-pecker, tawny owl and woodpigeon are found in the larger gardens from time to time, especially in those on the periphery of the built-up area.

Here is a typical report on the birds of a small suburban garden :

" The most abundant visitors to my garden are sparrows and starlings. I think there is but one pair of chaffinches in the road. The blue tit is but a rare visitor and the only tit I have seen since 1933. The blackbird is common and the singing posts are the chimneys. Occasionally only do I hear the tawny owl. The thrush is less common than the blackbird. My estimate for our road this year is three pairs of blackbirds, one pair of thrushes, one pair of chaffinches and one pair of ring-doves. A pair of carrion-crows were in the road last year, but I saw no proof of young. This year they have one young bird with them. The young bird takes his station between the chimney pots. There are four short pots to each shaft and the young bird is central. Old birds bring food to it. We have no robin or wren. I occasionally see a hedge-sparrow. The only time I have seen skylark or greenfinch was during the cold spell of December last (1938). There are about twenty houses on each side of the road. The back gardens are about 15 x 10 yards and meet the back garden of the next road. The gardens have each one or more fruit, &c., trees. Front gardens are almost negligible. There are plane, &c., trees in the road."

(Dr. J. S. Carter, Golders Green, 1939).

Farther out, at Richmond, Surrey, two full lists of the birds seen in small gardens during one year by Mr. C. L. Collenette in 1938 and by Major J. C. Eales-White in 1939 include twenty-four species that alighted in the gardens and nine more seen flying over, viz. :

Alighted in Garden

Carrion-Crow.	Nuthatch.	Blackbird.
Jackdaw	Great Tit.	Robin.
Jay.	Blue Tit.	Hedge-Sparrow.
Starling.	Coal-Tit.	Wren.
Greenfinch.	Long-tailed Tit.	Great Spotted Woodpecker.
Chaffinch.	Chiffchaff.	Lesser Spotted Woodpecker.
House-Sparrow.	Mistle-Thrush.	Tawny Owl.
Pied Wagtail.	Song-Thrush.	Woodpigeon.

Flew over Garden

Swallow.	Heron.	Black-headed Gull.
Swift.	Mute Swan.	Common Gull.
Kestrel.	Lapwing.	Lesser Black-backed Gull.

It is clear that all the birds that have adapted themselves to a greater or less extent to life in a suburban garden must have done so within comparatively recent times, for the simple reason that it is only within comparatively recent times that there have been any suburban gardens for them to adapt themselves to. The Continental races of many of our common garden birds have not yet or only recently adapted themselves to a garden habitat. In parts of Europe the robin, for instance, is known as a very shy bird of deep woodland. In the Mediterranean region the local race of the hedge-sparrow is said to be confined to mountain woods and scrub in the breeding season. The Continental song-thrush also has only within the past ten or fifteen years become a frequent inhabitant of parks and gardens in Germany, while in the south of its range it is confined, like the hedge-sparrow, to mountain woods. It is of interest to note that in New Zealand the garden association of birds consists almost entirely of species introduced from England, nearly all the native birds having failed to adapt themselves.

The chief condition of successful adaptation by a bird to life in fairly close association with man is that it should combine a certain degree of mistrust (and so avoid the fate of the great auk, the solitaire

and the dodo) with a large measure of familiarity, so that it can come
and take worms from under the spade like the robin, take crumbs from
the human hand like the house-sparrow (Plate XVa), or feed from a
coconut (Plate XIVb) or bird-table outside the window of an inhabited
room, like tits and starlings. It must also be able to rear its brood in
close proximity to human beings strolling up and down and making
a certain amount of noise. If birds can get used to the presence of
man, without becoming fatally friendly, then man will provide the
food, both wittingly and unwittingly. For man cultivates the soil, and
so multiplies manifold the amount of insect and other invertebrate
food that a given area of land will produce, not to mention the vegetable food—seeds, fruit and leaves.

The birds that become really common in suburban gardens are in
fact those that are bold enough to come to a bird-table, for they are
the ones which can breed fearlessly near man. Which birds actually
occur in any particular garden will be determined mainly by the
presence of suitable nesting-sites, such as shrubs and natural or artificial
holes, and the absence of enemies, such as cats and egg-collecting boys.
If nesting sites are short, but food supply plentiful, birds will nest in
some very queer places (Plates XIVa, XVIa). The robin's habit of
building its nest in old kettles or boots is well known (Plate XVII) and
one has been known to construct a nest in a gardener's jacket pocket,
hanging up in a shed, between the owner's breakfast and midday meal
on the same day. David Lack quotes the case of a pair of robins which
reared a brood in a garage in a London suburb, and made nuisances
of themselves by attacking human beings who approached.

The food of the garden birds leads us on to the next section of the
garden fauna. The list of insects and other invertebrates that have
adapted themselves to garden life is far too numerous to give even
one-half of 1 per cent of it here. It is only necessary to think of the
cabbages shredded to bits by the larvæ of the cabbage white butterflies
(*Pieris brassicæ, P. rapæ*) (Plates 15, 16a) and the cabbage moth (*Barathra
brassicæ*), of the broad beans infested by black fly (Plate XVIIIa), of
the roses covered with aphides which are themselves preyed on by
the larvæ of ladybirds, of the white-fly that invades our tomato-filled
greenhouses, of the wireworms and leatherjackets under the lawns, and
of the maggots of the codlin moths eating away inside the apples, to
realise with what ready adaptability the insect world has taken
advantage of man's horticultural activities. Man as a gardener

should be devoutly thankful for the birds that have chosen the role of insect-eater in his gardens. It is a partnership that should redound to the mutual advantage of both parties—provided the senior partner nets his cherries and raspberries.

Not all the insects that frequent gardens, however, are harmful. The dainty holly blue butterfly (*Lycænopsis argiolus*), whose larvæ feed on the buds and leaves of the ivy and holly, is common in London gardens and cannot be regarded as anything but an asset. The handsome small tortoiseshell butterfly (*Aglais urticæ*), though not exclusively a garden butterfly, is found wherever a patch of nettles provides sustenance for its caterpillars, and in August and September the buddleia bushes are usually covered with a mass of butterflies, large and small whites, small tortoiseshells, peacock butterflies (*Nymphalis io*), and often red admirals (*Vanessa atalanta*) (Plate 16a) and painted ladies (*V. cardui*) as well.

The late R. W. Robbins has left an interesting picture of the butterflies and moths to be found in his garden in South Hackney in the '80's. He found that a London garden provided a specialised habitat for its insect population, so that it maintained a small group of species that were uncommon in more rural districts. There was shelter from high winds and extremes of temperature, brick walls giving the protection and warmth of rocks without the heavy rainfall common in most rocky and wooded districts. Predators in the shape of insectivorous birds were relatively few (due in part to the large cat population ; cf. Darwin's famous chain of cause and effect linking the number of cats in a district with the prevalence of the red clover (*Trifolium pratense*) via field-mice and humble-bees).

"Brindled Beauties (*hirtaria*)[1] were common in the garden and doubtless bred on a small lime-tree in the front. But *hirtaria* was then of small account to me beside the Gothic (*typica*), which turned up occasionally on the walls. A species frequently taken was the small Angleshades (*lucipara*), the larvæ of which devastated our ferns, while the currant bushes were attacked by the Currant Clearwing (*tipuliformis*), which was to be found sunning itself on the leaves.

"With my net at dusk I used to take the Rustic Shoulder Knot

[1][The author here adopts the (to other zoologists) rather irritating entomological convention of giving only the specific names.—R.F.]

(*basilinea*), which I have hardly seen since ; the Lychnis (*capsincola*) on the pinks ; the Nutmeg (*chenopodii*), Garden Carpet, Brimstone, three or four Pugs (*assimilata, vulgata, centaureata*), and other well-distributed species. One well-known London insect, the Double-lobed (*ophiogramma*), I never found at Hackney, although we had in the garden the Ribbon Grass (*Phalaris*), in the stems of which the larva feeds. The moth was common at Clapton and Forest Hill."

As the food of birds leads on to insects, so the food of insects leads to plants. Most of the plants which provide food for the innumerable insects of the garden are man's imported and cultivated ones, roses, nasturtiums, cabbages, turnips and the like. There is, however, a very distinct weed flora of London gardens, consisting of wild plants that have adapted themselves to the easy life of a seed-bed or flower-bed, where the competition of other plants is kindly removed by man's digging and hoeing. (Plates, 17, Xb.)

Members of this association include such familiar enemies of the gardener as the shepherd's purse (*Capsella bursa-pastoris*), common chickweed (*Stellaria media*), creeping cinquefoil (*Potentilla reptans*), fool's parsley (*Aethusa cynapium*), ground elder or goutweed (*Aegopodium podagraria*), groundsel (*Senecio vulgaris*), sow-thistles (*Sonchus oleraceus* ; *S. asper*), dandelion (*Taraxacum officinale*), bindweed (*Convolvulus arvensis*), black nightshade (*Solanum nigrum*), speedwell (*Veronica persica*), red dead nettle (*Lamium purpureum*), greater plantain (*Plantago major*), white goosefoot or fat-hen (*Chenopodium album*), common orache (*Atriplex patula*), several docks (*Rumex crispus, R. obtusifolius, R. nemorosus*), sheep's sorrel (*R. acetosella*), polygonums (*P. aviculare, P. convolvulus, P. persicaria*) and spurges (*Euphorbia helioscopia, E. peplus, E. exigua*), not to mention the ubiquitous grass *Poa annua*.

Another specialised habitat is the mown lawn. Here there is a premium on plants that have flat rosettes of leaves, and can seed without raising their heads high enough to be cut off by the mower. The daisy (*Bellis perennis*) is rather good at this, and so survives longer than most of the plants that arrive already embedded in the turf, much to the gardener's annoyance. The dandelion and the ribwort plantain (*Plantago lanceolata*) are two other typical lawn plants that are ruthlessly eradicated by gardeners.

Many plants have been added to the flora of London as a result of escaping from the flower or vegetable garden. Escapes from the flower garden include the wallflower (*Cheiranthus cheiri*), the spur valerian (*Kentranthus ruber*), which is especially abundant on chalk cuttings in the lower Darent valley, the Michaelmas daisy (*Aster novi-belgii*), and the snapdragon (*Antirrhinum majus*), which grows on railway cuttings at Clapham Junction and Sutton, Surrey. From the vegetable garden have come such strays as the horseradish (*Cochlearia armoracia*), the garden cress (*Lepidium sativum*), angelica (*Archangelica officinalis*), which is cultivated for confectionery and has established itself along the Thames banks between Barnes and Kingston, and the mints *Mentha viridis*, associated with roast lamb, and *M. piperita*, associated with sweetmeats.

Turning now to the parks, we find that all the animals and plants of the garden association are found there also, only more so, so to speak. A special Committee has been set up by the Government to report annually on the birds of the Royal Parks (Hyde, Green, St. James's, Regent's, Richmond, Greenwich, Bushy and Hampton Court Parks and Kensington and Kew Gardens), so that we have very exact knowledge of the birds that breed in them. Practically all the birds mentioned above as belonging to the garden association breed annually in one or other of the Central Parks, and a surprisingly large number of others are regular passage migrants and winter visitors. The birds that nest annually include the carrion-crow, jackdaw (Kensington Gardens only), starling, greenfinch, chaffinch, house-sparrow, great and blue tits, spotted flycatcher (in both Hyde and Regent's Parks), mistle and song-thrushes, blackbird, robin, hedge-sparrow, tawny owl, mute swan, mallard, tufted duck, woodpigeon, moorhen and coot. In addition, a number of other birds, such as the jay, wren, great spotted woodpecker, cuckoo and dabchick nest from time to time.

The list of regular visitors to the Central Parks is most imposing. The passage migrants seen almost every year include the yellow wag-tail, chiffchaff, willow and garden warblers, blackcap, swallow, house and sand-martins, swift, common sandpiper and (most remarkable of all) woodcock. Non-breeding visitors at other times, usually winter, include, besides the gulls, such un-urban species as the brambling (in most years on Primrose Hill, just outside Regent's Park), the grey wagtail, tree-creeper, fieldfare, redwing (these two whenever there is any hard weather), kingfisher (especially on the canal in Regent's Park), lesser spotted woodpecker, kestrel, sparrow-hawk (usually a wanderer from Hampstead Heath to Regent's Park), heron (an attempt to nest on the roof of one of the Zoo aviaries was made in 1936), and great crested grebe. The list of rarities that have turned up from time to time in the London parks is still more striking ; it includes the hooded crow, wood-lark, merlin, Slavonian and black-necked grebes, red-throated diver, kittiwake, Iceland and little gulls, and Arctic skua (*Stercorarius parasiticus*).

A good idea of the normal winter bird population of the Central Parks is given by the census of Kensington Gardens taken by E. M. Nicholson in November, 1925. The most numerous bird in the gardens was, as might have been expected, the house-sparrow, with 2603 representatives, followed by 411 starlings, 289 black-headed gulls, 241 woodpigeons and 240 mallard. There was a considerable gap between these five abundant species and the next group, consisting of 37 blue tits, 26 moorhens, 21 blackbirds, 19 great tits, 16 robins and 16 pochard. Smaller numbers of carrion-crows, jackdaws, greenfinches, chaffinches, skylarks, coal-tits, mistle-thrushes, song-thrushes, hedge-sparrows, wrens and tufted ducks were also present, as well as single specimens of the pied wagtail, great spotted woodpecker and gadwall, making a total of 3981 individuals of 25 species, or about 14.5 birds per acre. On a later count in the same winter the brambling, redwing, stock-dove, common gull and coot were added to the list.

The two most spectacular examples of birds adapting themselves to the environment of the London parks are provided by the increase of the woodpigeon and the moorhen. The woodpigeon is so shy a bird in the country that it has always amazed countrymen to find it so tame in London. In the Committee's Report for 1937, Mr. Bayne quoted the case of a Scots gamekeeper who was asked, on returning home from a holiday in London, what had impressed him most about the city, and replied, " Seeing the cushie-doos in St. James's Park eating from a man's hand." There have probably always been a few woodpigeons in the London parks ; we know they were there in 1834. As late as 1886, however, there were hardly a dozen pairs nesting in Inner London. In the ensuing six to ten years, which also curiously enough saw the arrival of the gulls and the starlings, the woodpigeon increased rapidly, so that by July 1892 a flock of eighty-three, half of them young birds, could be seen in St. James's Park. At the same time its habits changed ; it became confiding and would take food from human hands. It also began to nest on buildings as well as in trees, though this tendency did not develop as it was once thought it might into a regular practice. To-day, the woodpigeon is one of the commonest breeding birds of the Central Parks, and one or two pairs are to be found in almost all the London squares (Plate XVIb) ; in the winter flocks of several hundred gather to roost in the tree-tops in Kensington Gardens and elsewhere.

Moorhens also colonised London towards the end of the last century. Hudson, writing in 1898, said they were almost unknown twenty years previously, but were then as widely diffused as the wood-pigeon. By 1930 as many as a hundred moorhens could be seen in St. James's Park alone at the end of the breeding season, and large numbers congregate there also in the winter months. At one time it looked as if the dabchick would follow the moorhen's example, for in 1897 seven pairs nested in St. James's Park, but of recent years it has become merely a visitor, rarely staying to breed, though it may do so regularly in Buckingham Palace grounds nearby.

Quite a number of butterflies find their way to the Central Parks, being especially attracted by the magnificent displays of dahlias and other flowers that are a feature of the parks in peace-time. Among those observed in Hyde Park and Kensington Gardens in the three years 1936-38 were the brimstone (*Gonepteryx rhamni*), red admiral (*Vanessa atalanta*) (Plate 16a), painted lady (*V. cardui*), peacock (*Nymphalis io*), comma (*Polygonia c-album*) (Plate 16b), clouded yellow (*Colias croceus*), small tortoiseshell (*Aglais urticae*) and meadow-brown (*Maniola jurtina*). As many as 46 moths (not counting " micros ") have been listed for the same two parks, the most striking being three hawk-moths, the lime (*Mimas tiliæ*), poplar (*Amorpha populi*) and eyed (*Smerinthus ocellatus*). Hyde Park is a well-known locality for the somewhat local yellow-legged clearwing (*Sesia vespiformis*), and its congener the red-belted clearwing (*S. myopaeformis*), which is commoner in gardens and orchards round London than anywhere else in the British Isles, has also been found there. A fly (*Syrphus guttatus*) of which the Natural History Museum possessed only one specimen, was taken in Hyde Park in August, 1938. A sample of honey from an apiary near Kensington Gardens was found to have an after-taste reminiscent of cats, but on maturing acquired " a delicious rich muscatel flavour." These flavours were attributed by Dr. R. Melville to the bees having visited predominantly the tree of heaven (*Ailanthus altissima*), a common street tree in Kensington, together with a fair admixture of sweet chestnut

(*Castanea sativa*) and privet (*Ligustrum vulgare*). Most London honey comes from the lime-trees (*Tilia* spp.), but in some seasons privet predominates, and then the honey has a dark colour and a bitter after-taste.

Whereas the bird life of the London parks has probably been more intensively studied than that of any corresponding areas anywhere in the country, the vegetation has been surprisingly neglected, though it presents a number of interesting features. Few flowering plants have managed to survive the continual trampling of the grass swards, and the swards themselves are often composed of species which do not usually form a sward, such as the annual meadow grass (*Poa annua*) and wall barley (*Hordeum murinum*). Ryegrass (*Lolium perenne*) is another common grass of the parks, while patches of cocksfoot (*Dactylis glomerata*) are not uncommon. Most of the common weeds can be found growing in odd corners, and in wartime the flower-beds are perforce not so well kept, so that they too are weed-strewn. In little more than an hour on June 25, 1944, some fifty-two different plants were noted in the north-east corner of Regent's Park, including two buttercups (*Ranunculus acer*, *R. repens*), the white campion (*Lychnis alba*), bird's-foot trefoil (*Lotus corniculatus*), three willow-herbs (*Epilobium*), enchanter's nightshade (*Circæa lutetiana*), wild chervil (*Anthriscus sylvestris*), two bindweeds (*Convolvulus*), two nightshades (*Solanum*) and the hedge woundwort (*Stachys sylvatica*). The last-named was growing in the okapi's enclosure in the Zoo.

Farther out than the Central Parks is a series of large open spaces, which have only been engulfed in the sea of houses and gardens within the past twenty or thirty years. They include Hampstead Heath and Ken Wood in Middlesex (Plates, 1, 18, 22), Hackney Marshes in the Lea Valley, Wanstead Park and Flats in Essex, Greenwich Park in the Kentish part of the County of London, Dulwich Park and Woods, Wimbledon Common and Richmond Park in Surrey. All of these are favourite London playgrounds, and most of them have received special attention from London naturalists. Here we may take Hampstead Heath and Ken Wood, including Parliament Hill Fields, as typical of these large outer parks and commons, though of course they all have their own peculiarities. (Map 7.)

In addition to the birds which breed in the Central Parks a surprisingly large number of different kinds of birds nest regularly on or quite near Hampstead Heath, and many others are more or less regular visitors without staying to breed. More than fifty species have been observed from the windows of a flat overlooking Highgate Cemetery in 1942-44 alone, including such unlikely ones as the haw-finch, lesser redpoll, black redstart, sparrow-hawk and tufted duck, not to mention a cockatoo that escaped from a local aviary. The birds that still breed on or just off the Heath comprise some forty-seven species, including the magpie (a pair that have bred in recent years in Ken Wood are thought to be escapes from captivity), jay, hawfinch, goldfinch, lesser redpoll, linnet, bullfinch, pied wagtail, tree-creeper, nuthatch, chiffchaff, willow-warbler, garden-warbler, blackcap, white-throat, lesser whitethroat, wren, house-martin, swift, kingfisher (a pair has frequented the Highgate Ponds for some time, and in the spring of 1944 was seen engaged in a courtship flight over the lake in Ken Wood), all three woodpeckers, cuckoo, kestrel, sparrow-hawk, and turtle-dove. This is quite an imposing list for an area which is now completely cut off by urban development from the open country, though only twenty-five years ago it was in fact possible to walk straight into the country from Jack Straw's Castle.

Since Harting contributed a list of the birds of Hampstead to Lobley's *Hampstead Hill* in 1889, many changes have taken place in the birds of the district, but they have not all been on the debit side. Though the rook, skylark, stonechat, reed-bunting, pheasant, sand-martin, nightingale, redstart, wood-warbler and a good many others have been lost as breeding species, such birds as the bullfinch, jay, green and great spotted woodpeckers, woodpigeon and moorhen, which were then rare have either become common or at any rate have not decreased further. The lesser redpoll, which in the '80's was known only as a rather unusual winter visitor, is now to be seen regularly throughout the breeding season, and its rather harsh flight-note, " chuch-uch-uch " can often be heard over the West Heath. It is interesting to note that in 1881 the quail nested in some fields near Parliament Hill.

The winter bird population of the Heath and its ponds is also of some interest. Redwings, meadow-pipits, and less frequently fieldfares are to be found, while the Highgate ponds, as Dr. Huxley and Mr. Best showed in their census in 1934, have a population of waterfowl that displays some remarkable fluctuations. The mallard, tufted duck, moorhen, coot and common and black-headed gulls are regular in the winter, but the pochard has forsaken the Highgate ponds since 1934.

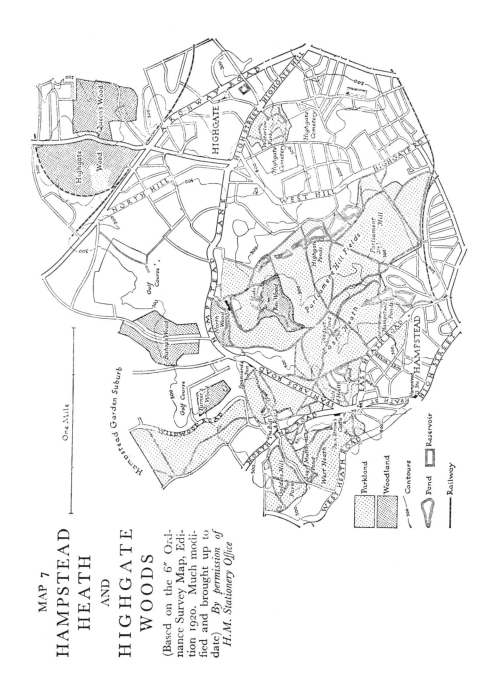

MAP 7

HAMPSTEAD
HEATH
AND
HIGHGATE
WOODS

(Based on the 6″ Ord-
nance Survey Map, Edi-
tion 1920. Much modi-
fied and brought up to
date) *By permission of
H.M. Stationery Office*

One Mile

Parkland
Woodland
300 Contours
Pond Reservoir
Railway

145

Great crested and little grebes are very occasional visitors, but herons can quite often be seen in the early morning.

Quite a number of butterflies are still to be seen on the Heath, On May 28, 1944, for instance, there were six different species, the three common whites, the wall brown (*Pararge megera*), small heath (*Cænonympha pamphilus*) and common blue (*Polyommatus icarus*). Just how good a hunting-ground for entomologists the Heath and its environs still form has been shown by Dr. I. H. H. Yarrow and Mr. K. M. Guichard, who have collected there within the past few years two-fifths of the known British species of the hymenopterous genera *Andrena* (host) and *Nomada* (parasite). Altogether twenty-six species of these burrowing bees, together with eleven species of their parasites are still to be found, while within the past hundred years six species of *Andrena* and seven species of *Nomada* have been lost.

The flora of the parts of the Heath to which the public has access has suffered somewhat from their attentions, but wherever a part of the Heath has been railed off, many very interesting plants still grow. Around the first of the Highgate Ponds, for instance, such plants as the lesser spearwort (*Ranunculus flammula*), marsh marigold (*Caltha palustris*), marsh-wort (*Apium nodiflorum*), angelica (*Angelica sylvestris*), bur-marigold (*Bidens tripartita*), water ragwort (*Senecio aquaticus*), gipsywort (*Lycopus europæus*), water mint (*Mentha aquatica*), skullcap (*Scutellaria galericulata*), great water dock (*Rumex hydrolapathum*), branched bur-reed (*Sparganium ramosum*), water-plantain (*Alisma plantago-aquatica*), yellow flag (*Iris pseudacorus*) and giant horse-tail (*Equisetum telmateia*) can still be found. In Turner's Wood, one of the fragments of woodland that still survive on the northern slopes of the Hampstead-Highgate ridge, though now quite surrounded by large houses with gardens, some typical woodland species still survive. They include the red campion (*Lychnis dioica*), holly (*Ilex aquifolium*), raspberry (*Rubus idæus*), dog-rose (*Rosa canina*), mountain ash (*Pyrus aucuparia*), angelica, guelder-rose (*Viburnum opulus*), figwort (*Scrophularia nodosa*), hornbeam (*Carpinus betulus*), and bracken (*Pteridium aquilinum*). This wood has been used for civil defence purposes during the war, and it is interesting to note that the frequent trampling down of part of the wood has brought in its train a number of weeds which are not normally found in woods, such as the hedge-mustard (*Sisymbrium officinale*), common chickweed (*Stellaria media*) and ribwort plantain (*Plantago lanceolata*).

Among other plants that still grow on the Heath may be mentioned the yellow water-lily (*Nuphar luteum*), called brandy-bottle from the smell of its flowers, which grows in the Ken Wood lake, the wood anemone (*Anemone nemorosa*), white bedstraw (*Galium mollugo*), devil's-bit scabious (*Scabiosa succisa*), so-called from the shape of its root, which the Devil is alleged to have bitten off, black knapweed (*Centaurea nigra*), water forget-me-not (*Myosotis scorpioides*) and water-pepper (*Polygonum hydropiper*). The most famous of all the plants of Hampstead Heath, the may-lily (*Maianthemum bifolium*) that used to grow, possibly native, in Ken Wood, became extinct, allegedly owing to careless path-making, but an attempt to reintroduce it a year or two before the war seems to have been successful, as it was still present in 1944.

References for Chapter 9

Bishop *et al.* (1928-36), Collenette (1939), Committee on Bird Sanctuaries in Royal Parks (England) (1929-39), Darwin (1859), Glegg (1935), Hampstead Scientific Society (1913), Harting (1889), Hudson (1898), Huxley & Best (1934), Johnson (1930), Lack (1943), London Natural History Society (1900-44), Melville (1944, 1945), Nicholson (1926a), Robbins (1939) Witherby *et al.* (1938-41), Yarrow (1941).

THE EFFECTS OF DIGGING FOR BUILDING MATERIALS

ONE of the indirect effects of the spread of buildings round London on the animal and plant communities has been due to the digging of building materials. Throughout the Home Counties, especially on the gravels and brick-earths in the valleys of the Thames, Lea and Colne, there are pits and hummocks due to this digging. Many of the pits are flooded, and so provide suitable haunts for aquatic birds and plants (Plates 2 and 3).

Already by the '60's there were complaints that the removal of brick-earth for the manufacture of bricks in the Heston district was constantly changing the level of the ground, and osiers (*Salix viminalis*) were sometimes planted in the resultant pits. At about the same time the Lord of the Manor of Hampstead was asserting his un-limited right to dig and carry away sand from Hampstead Heath, to the extent of destroying its herbage. According to Lord Eversley this digging was in fact being carried out to a degree that threatened to interfere with the natural features of the Heath, as dangerous pits were appearing in many places and cutting up the surface of the ground. This can be verified by anybody who takes a walk on the East Heath to-day. Near the Vale of Health the ground surface is very uneven, even more so since the digging of sand to fill sand-bags in September, 1939. The Lord of the Manor's threat to the amenities of the Heath was one of the main causes of the successful agitation for the public acquisition of this most popular of London's larger open spaces. The East and West Heaths were bought by the Metropolitan Board of Works in 1869, and to this nucleus was added Parliament Hill Fields in 1889, Golder's Hill Park in 1898, the extension towards Hampstead Garden Suburb in 1905, and finally Ken Wood and the adjoining fields in 1925.

In other places round London the threat to the amenities of the

PLATE 27

Hainault Forest, Essex, from Dog Kennel Hill

The whole of this area was ploughed up a hundred years ago   ERIC HOSKING

Daffodils at Ken Wood   ERIC HOSKING

Crocuses at Hyde Park Corner   ERIC HOSKING

open spaces came from the commoners rather than the Lord of the Manor. In Epping Forest gravel and sand-pits were dug all over the Forest, the materials being used without restriction, while turf for suburban villa lawns was stripped from large areas. At Wimbledon both commoners and lord dug gravel, loam and peat on the common, the lord finally establishing his right to do so without limit.

The gravelly parts of the London area, particularly in Western Middlesex, are studded with gravel-pits, both used and disused, that have filled with water, and so increased manifold the available habitats for aquatic animals and plants. Such pits may be found, for example, at Harefield (Plates 2 and 3), Hatton, Feltham and Ashford in Middlesex, Rickmansworth, Hamper Mill and Cheshunt in Herts, Sewardstone and Dagenham in Essex, the Cray and Darent valleys in Kent, Beddington, Mitcham Junction and Walton-on-Thames in Surrey, and Colnbrook in Bucks. Their vegetation includes such water-loving species as the flowering rush (*Butomus umbellatus*), bur-marigold (*Bidens cernua, B. tripartita*), codlins-and-cream (*Epilobium hirsutum*), great reed-mace (*Typha latifolia*), often called bulrush on account of the botanical ignorance of a Victorian artist who erroneously painted Moses's cradle surrounded by this species, and the common reed (*Phragmites communis*), the favourite nesting-site of the reed-warbler. The real bulrush (*Scirpus lacustris*), which is actually a sedge, is rather local round London, especially south of the Thames.

Since the banks of the flooded gravel pits provide so much cover, and there are often small islands as well, many more birds breed on them than on the concrete-banked reservoirs of the Metropolitan Water Board. They include reed-buntings, sedge and reed-warblers, moorhens, coots, mallards, dabchicks and great crested grebes. At the census of great crested grebes taken in the London area in 1935, nesting pairs were found on six gravel pits, Colnbrook, Ashford, Feltham, Hamper Mill, Sewardstone and Mitcham Junction, while three other pits, Hatton, Beddington and Walton-on-Thames, held non-breeding pairs. Altogether 38 per cent of the breeding and 36 per cent of the non-breeding great crested grebes in the London area were found on flooded gravel pits in that year.

Typical winter populations of the waterfowl on the gravel-pits round London are the following from the census of certain waterfowl (swans, mallard, dabchicks and moorhens excluded) taken in the London area on December 18, 1937 :

Beddington Lane—6 Pochard, 1 Tufted Duck, 65 Coot.
Cray Valley—21 Pochard.
Hamper Mill—10 Teal, 20 Pochard, 14 Tufted Duck, 2 Great Crested
   Grebes, 78 Coot.
Mitcham Junction—56 Pochard, 30 Tufted Duck, 150 Coot.

Usually it is only the commoner ducks, mallard, pochard and tufted, and sometimes teal, that turn up at the gravel pits, the rarer ones preferring the reservoirs, which are larger. During 1938, however, wigeon were reported from Beddington Lane and Hamper Mill, shoveler from Mitcham Junction, goldeneye from Cheshunt and smew from Beddington.

In 1944 one of the most interesting events in the history of the avifauna of the London area took place when a pair of little ringed plover (*Charadrius dubius curonicus*) nested on a shingle-bank at the edge of a large gravel-pit in South-west Middlesex. Prior to 1944 the species had only once nested in the British Isles, at Tring Reservoirs, Hertfordshire, in 1938, and in 1944 two pairs also nested at Tring. This bird favours shingle-banks in rivers for its nesting-sites on the Continent, where it breeds regularly as near to our shores as Holland.

One other bird species is closely linked with the distribution of sand and gravel-pits, and that is the sand-martin, which nests in colonies in the small artificial cliffs formed in the pits. Sand-martins no longer nest so near the centre of London as in Gilbert White's day, for their possible nesting-sites are steadily filled in and built over as London spreads. Nevertheless there are still plenty of sand-martins on the outskirts of London. Indeed in the London area the bird is almost completely dependent on man for its nesting sites, only one completely natural site having been recorded in recent years, in a sandy bank on the Thames near Shepperton. Of the forty-eight other sites recorded between 1900 and 1940, eighteen were in gravel or ballast pits, thirteen in sand-pits, five each in railway cuttings or embankments and in drain-pipes, three in holes in the walls of wharves, and one each in a canal-bank, a road-cutting, brick-work and trenches. Few birds so completely adapt themselves to man-made habitats as the sand-martin has near London. As a result its distribution in the London area is confined to districts where the subsoil is sand, gravel, brick-earth or alluvium.

Since sand-martins nest in pits that are often still being worked, their colonies tend to be rather ephemeral, though Plate 4, taken in a disused sand-pit on the Barnet By-pass now converted to a rubbish-dump, shows how tenaciously they will cling to a site so long as the cliff-face is not disturbed. Some colonies are driven out by the whole pit being filled up in this way. When suitable pits are lacking, sand-martins readily turn to other kinds of hole. In 1923 several sand-martins were observed carrying nesting material into holes in the brick-work of the platform at Rye House near Hertford, though the station was being regularly used by trains. Equal adaptability was shown by the birds lower down the Lea valley when they forsook the local ballast pit to nest in the drain-pipes of the aqueduct that runs southwards from King George V Reservoir, and by others that nested in 1917 in trenches dug for military training at Gidea Park, Essex.

The nearest sand-martin colony to London at present is probably the one near Earlsfield Station, five miles from Waterloo on the main Bournemouth line, but within the past twenty years there have also been colonies at Ken Wood and Hurlingham, five and four miles respectively from Charing Cross.

References for Chapter 10

Eversley (1910), Fitter (1941b), Hampstead Scientific Society (1913), Hollom (1936), Homes (1938), London Natural History Society (1937-44).

THE INFLUENCE OF TRADE AND TRAFFIC

FOR many centuries now London has been one of the world's great ports. It was as a port that London's fame originally spread, and it is to the sheltered anchorage provided by the queenly Thames that the City has above all owed its great position, and hence its teeming population. It is therefore of some interest to examine the extent to which the fact that London is a great centre of trade and traffic has influenced its animal and plant communities.

Many unwelcome animal and plant visitors, the cockroach, the bed-bug and the black and brown rats, to name only a few, have come to London accidentally in the course of trade. To facilitate trade, the Thames has been bridged and embanked, its low-lying flats have been excavated to form docks, and engineers have even spoken of throwing a great barrage across its mouth. To facilitate trade, roads have been made, railways driven across the clayey countryside, and canals with their feeder reservoirs constructed.

Mention has been made already of some of the unpleasant aliens that have come to Britain through London and other ports in the course of commerce. Among those that have not succeeded in establishing themselves on such a scale as to be national pests, may be mentioned such " camp-followers of commerce," as Professor Ritchie calls them, as the lentil beetle (*Bruchus lentis*), which has been noted among Egyptian seeds at Gravesend, the Chinese lentil beetle (*B. chinensis*), found in imported lentils in several London suburbs, and the larvæ of another beetle, *Lyctus brunneus*, which have been taken from wood used by a London taxidermist for mounting specimens.

The insect fauna of London warehouses, mainly introduced, is both plentiful and varied. Every commodity has its own special pests. Flour alone has two beetles and three moths among its commonest predators. A mite, *Pediculoides ventricosus*, which normally eats the

Plate XVII

Robin's nest in old watering-can   ERIC HOSKING

Plate XVIII

An aspidistra in a London parlour JOHN MARKHAM

Black fly on broad bean S. BEAUFOY

grubs that eat the grain, is not above transferring its attentions to the men that eat the grain products as well. When it does, an extremely irritant itch results, so that when in 1913 several cargoes of grub-infested cotton-seed from Egypt arrived in the Port of London, the dockers' wages had to be raised by 50 per cent to induce them to unload these ships. Sugar has its mite (*Glycophagus*), tobacco its beetle (*Lasioderma serricorne*), and peas another beetle (*Bruchus pisorum*). The fauna of biscuits is extraordinarily rich. An analysis of tins of army biscuits in 1913, for instance, yielded four species of moth, twelve beetles and a hymenopterous insect.

The Port of London abounded with rats and mice until the intensive extermination campaign of the last three years. In 1942 about one-quarter of all the premises in the port, mainly warehouses, were known to contain at least one rat, including 21 per cent inhabited by black rats and 16 per cent by brown rats. (Plate VI a.) Many buildings thus held both kinds of rat, and there was also an overlap between rats and mice, which were found in three-fifths of all infested premises in the port. Out of 1020 rats killed in the Port of London in 1941-42, 81 per cent were black and the remainder brown. Recent scientific investigations into methods of rat control in the Port of London by Morgan and others have shown that the cats which are kept on the pay-rolls of many warehouses to kill the rats are usually far too well fed to carry out their prime function, being as often as not on very good terms with their " prey."

Three forms of the black rat occur in the Port of London ; they are often set down as sub-species, but are probably mere colour-phases. In addition to the normal black form (*Rattus rattus rattus*), there are the Alexandrine rat (*Rattus r. alexandrinus*) and the tree or roof rat (*Rattus r. frugivorus*). The roof rat has long dense fur, light brown or grey above and white or yellowish below, and its feet are usually white above. The Alexandrine rat can be told by its brown back and dusky underside, the sharp nose and long tail distinguishing it from the lumbering, blunt-nosed, short-tailed brown rat (*Rattus norvegicus*). To add to the confusion there is a rare black form of the brown rat, which is distinguished from the normal black rat by its blunt nose and short tail.

A most interesting example of adaptation is to be found in the cold stores in the Royal Albert Dock, where brown rats and house-mice flourish and sometimes breed at a temperature of 17°F. by dint of growing slightly thicker fur and layers of fat underneath. The house-mice dine off bags of frozen kidneys, in which they make their nests and rear their naked young. The black rat has never been found in the refrigerated chambers, but inhabits the top floors, where the machinery is kept and the temperature is normal.

The outstanding feature of the modern Port of London (Plates 6b, XIX) is its extensive system of docks, covering in all many hundreds of acres of water, which were constructed from the beginning of the nineteenth century onwards, as shown in the following table :

Name of Dock	Date of First Opening	Water Area, 1927 acres
West India	1802	$92\frac{1}{2}$
London	1805	$35\frac{1}{2}$
East India	1806	$31\frac{1}{2}$
Surrey	1807	147
St. Katherine	1828	10
Royal Victoria	1855	93
Millwall	1868	35
Royal Albert	1880	87
Tilbury	1886	90
King George V.	1921	64

These large artificial lakes in the lower Thames valley are too much frequented by ships to make them such ideal resorts for waterfowl as the reservoirs we shall discuss in the next chapter. However, numbers of mallard and occasional flocks of tufted duck can be seen there, as well as mute swans, and a few years ago an escaped black swan was a regular sight in the docks along the Blackwall Reach.

The relative lack of birds, however, is compensated for by an abundance of fish. Mr. P. W. Horn, to whom we are indebted for most of what we know of the fishes of the London docks, has recorded thirteen species and one hybrid from the docks in the borough of Stepney. These fish appear to be resident and to breed, in spite of the poisonous state of the river water outside. This is due partly to the fact that the salinity of the dock water is comparatively constant, owing to the narrowness of the entrance and exits, and partly to the fact that stationary water stored in open reservoirs is purified both chemically and mechanically. In the docks the water is so clear that you can see objects six feet down, whereas in the river outside it is almost completely opaque.

Nearly a hundred years ago the London docks were a favourite resort of tradesmen anglers, who sometimes secured bags of 50 lb. of bream, perch and roach. Unfortunately a fatal accident to an angler resulted in the withdrawal of the privilege by the dock companies. Thereafter only hearsay evidence was available until 1921, when Mr. Horn, Curator of the Whitechapel Museum, was able to obtain specimens of the fish that were brought to the surface by the abnormal heat of that summer. The commonest species were the roach, dace, bleak (*Alburnus alburnus*), gudgeon, perch and three-spined stickleback (*Gasterosteus aculeatus*). There were also odd specimens of the bream (*Abramis brama*), minnow (*Phoxinus phoxinus*), goldfish (*Carassius auratus*, obviously escaped), pike and trout. The trout was picked up dying on the sill at the Shadwell entrance to the docks, having probably drifted down from the upper reaches of the Thames. A roach-bream hybrid was found quite commonly, which is curious in view of the fact that the only bream recorded was a diseased one in April, 1923.

The condition of the fish from the docks of Stepney compared very favourably with that of fish from other London waters. The external parasitic fish-louse *Argulus foliaceus*, which infests fish from the Regent's Canal, was not seen on any dock fish, though a few bruised and fungous fish were brought in. The mortality in the Stepney docks in the hot summer of 1921 was not so great as that in the Surrey Commercial Docks on the other side of the river, where the quantity of dead fish was estimated to run into tons. Mr. Horn attributed the abnormal sickness and mortality to the de-oxygenation of the water and the generation and liberation of foul gases from the bottom, combined with excessive salinity due to the high tides prevailing against the diminished flow of fresh water from the drought-ridden upper Thames valley.

The fish appear to have plenty to feed on after they have passed the stage of eating infusoria. At the end of May water-fleas (*Daphnia major*) appear in the quieter corners of the docks in such enormous numbers as to colour the water, and both old and young fish devour them avidly. A dense growth of silky green algæ coats the piles and woodwork, and around these Mr. Horn saw perch and roach browsing.

From the docks it is only a short step to the canals which terminate in them. The Regent's Canal (Plate IV) joins the Grand Union

MAP 8 The County of London, showing the Metropolitan Boroughs. (Based on the Diagram of London,

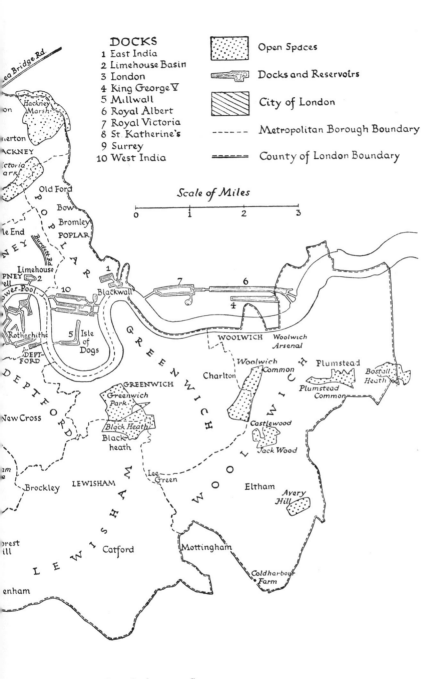

DOCKS

1 East India
2 Limehouse Basin
3 London
4 King George V
5 Millwall
6 Royal Albert
7 Royal Victoria
8 St. Katherine's
9 Surrey
10 West India

Open Spaces

Docks and Reservoirs

City of London

------ Metropolitan Borough Boundary

===== County of London Boundary

Scale of Miles

0   1   2   3

published by the Ordnance Survey, 1935) *By permission of H.M.
Stationery Office*

157

Canal to the Limehouse Basin, between the London and West India Docks. The Grand Union Canal also links on to the Brent Navigation, with its mouth at Brentford, before passing on to Uxbridge, and up the Colne and Gade valleys to the Midlands and the north. (Plate V). The Grand Surrey Canal leads from the Surrey Docks into the heart of South London, to Peckham and Camberwell. Formerly the Surrey Docks were also joined to Croydon by another canal, whose track is now occupied by the main railway line from London Bridge to Croydon.

The fauna of the canals is very similiar to that of the docks. Ducks, especially mallard, may be seen at many points along them. At St. Mary's Basin behind Paddington Station, where the Regent's and Grand Union Canals join, tufted duck may sometimes be seen in the winter. Kingfishers, herons and grey wagtails are among the more unexpected visitors to the quieter stretches of the Regent's Canal.

Far more important than the canals themselves as resorts of bird and fish life are the three canal reservoirs in the London area, Aldenham or Elstree (Plate 10a ; 65 acres), constructed in 1797, the Brent or " Welsh Harp " (197 acres), constructed in 1810 and enlarged in 1851, and Ruislip (43 acres), constructed in 1810. All three, together with the even more famous Tring Reservoirs in Hertfordshire, feed the Grand Union Canal, and are favourite haunts of both birds and bird-watchers. The Brent has always been the most productive of rare birds, but though its great days described in Chapter 5 are now over, not a few interesting species have been seen there within recent years ; for instance, two spotted redshanks on September 11, 1938, a red-necked grebe on December 27, 1940, upwards of seventy smew on February 28, 1943 and a turnstone on May 11, 1944. All three canal reservoirs near London are now much frequented by the public—Aldenham and Ruislip are used as boating lakes—so that they are not so attractive to birds as they once were.

Few birds now breed at the Brent Reservoir, as its banks have little vegetation and are well trodden by small boys. Nevertheless one or two pairs of yellow wagtails hang on, and one nest was found in 1943. Aldenham had the distinction of being the breeding-place of a pair of garganey in 1931—the only recent breeding record for this species for the London area—but otherwise only a few pairs of mallard, grebes, moorhens and coots carry on the unequal struggle against the boaters and anglers in the breeding season. Ruislip is still, in spite of its

" lido," the best of the three for birds in the spring and summer. Both sedge and grasshopper-warblers occur, and this is one of the very few places in the London area where teal have been known to breed within recent years. Here also a few moorhens, coots and crested grebes battle against the advancing legions of boaters. The grebes are particularly vulnerable to water-borne assault, for their nests are usually out on the open water, whereas the moorhens and coots seek the shelter of the marginal rushes. In the winter jack-snipe frequent the edge of the reservoir, and water-rails hide in the thick herbage at the northern end. The adjacent common is noted for its nesting red-backed shrikes, and is easily the best place round London to find a willow-tit between September and March.

Many interesting aquatic plants still grow on the margins of the canal reservoirs. At Aldenham or Ruislip such plants as the lesser spearwort (*Ranunculus flammula*), bur-marigold (*Bidens*), water forget-me-not (*Myosotis scorpioides*), flowering-rush (*Butomus umbellatus*) and the rare fox-tail grass *Alopecurus fulvus* can be seen. The Brent is less fruitful now, but is still known as a locality for the scarce dock *Rumex palustris*.

Whereas the influence of water transport on the natural communities is mainly a beneficent one, that of land transport, roads and railways, is largely lethal. Motor-cars, which Marie Lloyd wittily called " smelly sparrow-starvers," have replaced the horses on whose dung the sparrows fed, and kill thousands of small animals every year, especially on the more rural roads round London. Beadnell has analysed over 2000 vertebrates found dead on British roads, and found 81 per cent birds, 14 per cent rabbits, 4 per cent hedgehogs, with the balance consisting of rats, frogs, toads, moles, snakes, dogs, cats, two sheep and one forest pony. Altogether one dead animal was found every 4.3 miles of road over some 10,000 miles. Many insects also get sucked into the radiators of cars, which often look like a butcher-bird's larder after a journey on a summer's evening.

Railways, especially electric ones, are almost as dangerous to the local fauna as roads. Badgers are quite often killed by trains, and the London Passenger Transport Board once had a somewhat grisly exhibition at Charing Cross Station of the badgers, otters and foxes that had been run over by its trains. A badger was run over by a train near Theydon Bois, Essex, in September, 1888, two were electrocuted on the line at Worcester Park, Surrey, in July, 1925, and a third at

Norbiton, not far away, in 1927. In September, 1936, a female badger weighing 22½ lb. was found run over in Camp Road, Wimbledon Common, and is now in a glass case in the Wimbledon Common Golf Club-house. A veritable battue of badgers occurred in the Purley-Whyteleafe-Caterham district of Surrey early in 1940, six having been found dead on the railway lines within a fortnight. Otters, too, have suffered road and rail casualties around London. One was electrocuted on the railway line at Chiswick in 1923, and another was run over by a car at night in Blake Hall Road, Wanstead, in November, 1928.

The large areas covered by railway sidings and goods yards in and around London are comparatively barren of plant life, but are much frequented by sparrows, and have recently been increasingly favoured by gulls, both common and black-headed, which scavenge there. The sides of railway embankments, on the other hand, are sanctuaries of plant, and to a smaller extent of bird, life, sometimes extending quite near the centre of London, as by the main lines from London Bridge to Croydon between New Cross Gate and Brockley, and from Victoria to Croydon between Clapham Junction and Balham. It was of this latter line that Richard Jefferies wrote some sixty years ago :

"The smooth express to Brighton has scarcely, as it seems, left the metropolis when the banks of the railway become coloured with wild flowers. Seen for a moment in swiftly passing, they border the line like a continuous garden. Driven from the fields by plough and hoe, cast out from the pleasure-grounds of modern houses, pulled up and hurled over the wall to wither as accursed things, they have taken refuge on the embankment and the cutting.

"There they can flourish and ripen their seeds, little harassed even by the scythe and never by grazing cattle. So it happens that, extremes meeting, the wild flower, with its old-world associations, often grows most freely within a few feet of the wheels of the locomotive. Purple heathbells gleam from shrub-like bunches dotted along the slope ; purple knapweeds lower down in the grass ; blue scabious, yellow hawkweeds where the soil is thinner, and harebells on the very summit ; these are but a few upon which the eye lights while gliding by."

In addition to providing sanctuary, the railways act as a distributive agent for spreading the seeds of plants, which get sucked along

The Upper Pool from the south end of Tower Bridge. JOHN MARKHAM

Plate XIX

Plate XX

Chiswick Eyot from Chiswick Mall   JOHN MARKHAM

by the trains as they rush to and fro.[1] The yellow toadflax (*Linaria vulgaris*) and the garden snapdragon (*Antirrhinum majus*) are two common plants of London railway embankments that probably get distributed in this way, while the Oriental bladder senna (*Colutea arborescens*) has become firmly established by the North London line at Dalston and by the District line at Barking. These are plants that can be seen and identified from the train. It is a pity that London botanists have never done any field study on the plants of London's railway embankments ; at all events nothing on the subject appears to have been published.

References for Chapter 11

Beadnell (1937), Bishop *et al.* (1928-36), Durrant & Beveridge (1913), Glegg (1935), Horn (1923), Jefferies (1883), Johnson (1930), London Natural History Society (1900-44, 1937-44), Marshall (1944), Morgan *et al.* (1942, 1943), Owen (1927), Ritchie (1920).

[1] The underground railways, as recent research suggests, may be acting in a similar way with a man-biting mosquito, *Culex molestus*, which breeds in stagnant water in sumps and under platforms (Marshall, 1944).

THE INFLUENCE OF WATER SUPPLY

IT IS a far cry from the days when the Thames and its tributary rills and rivulets were the direct source of London's water supply, though well into the nineteenth century the intake pipe of one water company at Chelsea was situated only two or three yards from the outlet of a large sewer in the river. To-day an abundant supply of pure fresh water is provided by a public body, the Metropolitan Water Board, from a series of large reservoirs in the Thames and Lea valleys, supplemented to a small extent by wells in the chalk of the North Downs. The water in the reservoirs comes, of course, ultimately from the Thames or the Lea, but there is no chance at all of any of the flora or fauna of these rivers suddenly appearing in Londoners' Saturday night baths.

There is nevertheless a very extensive flora and fauna associated with reservoirs and waterworks. It may be divided into two classes, the birds, which live on the surface, and all the rest, which live in the water and water-pipes. This latter flora and fauna have been admirably described by Dr. Anna Hastings in a British Museum pamphlet.

The typical flora of a waterworks consists of algæ, fungi and bacteria. The three most important groups of algæ usually found in waterworks are the Isokontae or Green Algæ, the Myxophyceæ, or Blue-Green Algæ and the Diatoms. Most of them are microscopic, but others are quite visible to the naked eye, in various shapes, such as the green threads which can often be seen on ponds. They come in with the river water, and create serious problems of water purification. Once in the reservoirs, they are apt to multiply exceedingly, especially in spring and autumn. In the spring of 1936, for instance, a diatom, *Fragilaria crotonensis*, was present in one of the Thames reservoirs in such large numbers that when its growth reached a maximum in May it was calculated that the total dry weight of this alga in the reservoir was of the order of 110 tons, of which 63 tons would be silica.

Algæ can be killed by algicides such as potassium permanganate, but dead algæ are sometimes as undesirable as live ones, owing to putrefaction. Certain algæ, however, are not amenable to coagulant algicides, for instance *Oscillatoria rubescens*, which floats on the surface of the water and may contain a purple pigment. In 1931 large quantities of *rubescens* coloured the water in one of the Thames reservoirs like cocoa, and stained the sand of the filters a royal purple. This reservoir had to be put out of action until the plague died down, and the disposal of the water from it became a problem, as it could not be released into the Thames again.

Algæ often complicate the process of filtering the water, as they soon collect in a film on the sand. The small green algæ called *Chlamydomonas* are very active swimmers, being equipped with a pair of flagellae, and are thus able to work their way through the sand filter, to the alarm of those consumers who do not like their water to be even faintly tinged with green, however harmless the causative algæ.

Bacteria also find their way into the reservoirs, and like the algæ are nowadays effectively eliminated by filtration. The iron bacteria, *Crenothrix*, are among the more unpleasant, having an effect on water so similar to the alleged results of the first plague of Egypt (" And all the waters that were in the river turned to blood. And the fish that was in the river died ; and the river stank, and the Egyptians could not drink of the water of the river." *Exodus*, vii, 20-21), that it has been suggested that *Crenothrix* was the agent of the Lord on this occasion. Another iron bacterium, *Gallionella*, is less spectacular, but more persistently annoying, as it coats up the insides of water-pipes, and may seriously reduce their capacity.

The fauna of the waterworks is even more interesting than the flora, and ranges widely over the animal kingdom. Polyzoa, worms, sponges, molluscs, rotifers, crustacea and fishes may all be found from time to time. In Hamburg at the end of the last century a quite fantastic collection of animals roamed the water-pipes of the city, into which the River Elbe flowed unfiltered. Dr. Hastings remarks :

" So we have the astonishing picture of the water-pipes of a great city encrusted almost throughout with Sponges, Polyzoa and Molluscs, sheltering a dense population of Crustacea and Worms, and with many thousands of Eels swimming in the fairway."

Happily, London's water is perfectly well filtered, so we are spared the eels, but even so eleven species of molluscs were once taken from a Poplar water-main, and 90 tons of the introduced zebra mussel (*Dreissensia polymorpha*) were removed from a quarter-mile stretch of unfiltered water-main at Hampton, Middlesex, in 1912—the diameter of the pipe had been reduced from three feet to nine inches. The zebra mussel is a native of Europe, and was probably introduced here from the Volga on timber. It was first recorded in Britain in 1824, when large numbers were found attached to shells and timber in the Commercial Docks on the Thames and were sent to the Linnean Society. It has since colonised more than twenty English and Scottish counties. In London it has been found in the reservoir near York Way that supplies the Metropolitan Cattle Market in Islington.

Besides the zebra mussel the duck-mussel (*Anodonta anatina*) and fresh-water snails, such as *Limnæa auricularia*, are sometimes found in waterworks, but are much less troublesome.

The sponges that grow in waterworks and water-pipes are fresh-water ones, such as *Spongilla lacustris*, and they and the polyzoa look to the uninitiated more like plants than animals. The fresh-water hydroid *Cordylophora lacustris* also forms plant-like colonies, which consist of horny stems that have tentacle-bearing animals at their ends. Other members of this interesting fauna include oligochæte worms, flat-worms, thread-worms, freshwater shrimps, water wood-lice and the larvae of midges, such as *Chironomus*.

In 1884 the water-pipes of the East London Water Company were invaded by eels, some of them eighteen inches long. There was much speculation as to how they might have got there, but it was eventually found that they had wriggled up the shelving bank of one of the reservoirs in the Lea Valley and entered some ventilation holes.

Kräpelin, who investigated the luxuriant fauna of the Hamburg water system, sought, among other things, proof of his theory that the blind and colourless water fauna which inhabits certain dark caves and very deep lakes would either colonise or evolve in the very similar environment of water-pipes which are never reached by light. He did not find what he was looking for in Hamburg, owing, he surmised, to the fact that the pipes had only been laid down for about thirty years. More recently, however, a member of the Metropolitan Water Board's staff found one of these blind underground creatures,

PLATE 29

JOHN CLEGG

Tulips in St. James's Park

PLATE 30

L. JAMES

Horse-chestnuts in Bushy Park

a shrimp-like crustacean called *Niphargus*, apparently breeding in some culverts under a filter that had been in use for about thirty years.

The ecology of this waterworks fauna and flora is a fascinating field of study. To give only one example of interaction, the clearing of mussels from the walls of a reservoir has been followed by the appearance of an immense swarm of algæ, which the mussels had hitherto helped to keep down. The big habitat distinction is between the free-swimming fauna and flora and those, like sponges, snails, and sessile diatoms, that need some kind of solid support, either the sides or bottom of the reservoir. The mud that accumulates on the bottom of the reservoir provides another important habitat, for worms, duck-mussels, midge larvæ and similar lovers of and livers in mud.

The creation by the Metropolitan Water Board of an extensive series of large lakes in the valleys of the Thames and Lea has been one of the major events in the recent development of London's avifauna. These large sheets of water have made London one of the best centres for the study of aquatic birds in winter in the whole British Isles. In an average winter practically every duck, grebe and diver that visits the British Isles regularly may be seen with tolerable certainty on one of the reservoirs within twenty miles of London. Of the nineteen species of duck which occur regularly in Britain, only two did not turn up on one of the London reservoirs in the single year 1938. These were the eider (*Somateria mollissima*), which is largely sedentary and does not breed south of Northumberland on the east coast, and the velvet scoter, which did occur in 1937. Similarly, all five species of grebe, and two out of the three divers that occur regularly in Britain, were seen on one or other of the London reservoirs in 1938. The remaining diver, the black-throated, was seen in both 1937 and 1939. Thus in the two years 1937-38, of the twenty-seven British species of duck, grebe and diver, only one, the common eider, was not seen on a London reservoir. It is safe to say that no other inland area in Britain could surpass this record, and few could hope to equal it.

The Metropolitan Water Board has more than fifty individual reservoirs, but from an ornithological point of view only the larger ones are really important. These are listed here, together with their areas, average depth and date of construction :

Reservoir	Date Constructed	Area in Acres	Depth in Feet
LEA VALLEY (Essex and Middlesex) :			
King George V, Chingford -	1900-13	416	27-32
[1]Walthamstow (12 reservoirs) -	1853-97	449	10-34

[1] Plate 10b.

THAMES VALLEY (Middlesex) :			
[2]Stoke Newington (2) - - -	1834	$42\frac{1}{2}$	12-15
Staines (2) - - - -	1902	424	29-39
Kempton Park (2) - - -	1906	62	20
Queen Mary, Littleton - -	1925	723	38

[2] Plate 9.

THAMES VALLEY (Surrey)			
Barn Elms (4) - - - •	1898	$86\frac{1}{2}$	
Lonsdale Road, Barnes (3)		31	
Island Barn, Molesey - - •	1911	188	
West Molesey - • • •			

It was not till some time after their construction that bird-watchers began to pay attention to these reservoirs, and the county histories of Essex and Surrey, published in 1890 and 1900 respectively, hardly mentioned them. C. J. Cornish was one of the first naturalists to give them the attention that was their due, and he has left a graphic description of a visit to Barn Elms Reservoirs, which lie on the south side of the river between Hammersmith and Putney Bridges, in February, 1902 :

" The scene over the lakes was as sub-arctic and lacustrine as on any Finland pool, for the frost-fog hung over river and reservoirs, only just disclosing the long, flat lines of embankment, water and ice ; the barges floating down with the tide were powdered with frost and snow-flakes, and the only colour was the long, red smear across the ice of the western reservoir, beyond which the winter sun was setting into a bank of snow-clouds. It was four o'clock, and nothing apparently was moving, either on the ice or the water, not even a gull. In the centre of the north-eastern reservoir was what was apparently an acre of heaped up snow. On approaching nearer this acre of snow changed into a solid mass of gulls, all preparing to go to sleep. If there was one there were seven hundred, all packed together for warmth on the ice. . . . Beyond the gulls, which rose and circled high above in the fog with infinite clamour, were a number of black objects, which soon resolved themselves into the forms of duck and other fowl. Rather more than seventy were counted, swimming on the water near the bank or sitting on the ice. . . . The result of a look through the glasses was something of a surprise. They were not mallard, teal or widgeon ; but three-quarters of the number were tufted ducks. . . . Some were sleeping, some diving, and other swimming quietly. When approached, the whole flock rose at once, and flew with arrow-like speed round the lakes and twice or thrice back over the heads of their visitors, of whom they were not at all shy, being used to the sight of the man who keeps the reservoir's banks in order. They swept now overhead, now just above the ice, like a flock of sea-magpies or ice-duck playing before some North Atlantic gale. As several birds had not risen, we ventured still nearer, and saw that most of these were coots, some ten or eleven, which did not fly, but ran out on to the ice. Two large birds remaining, which had dived, then rose to the surface, and to our surprise and pleasure proved to be great crested grebes."

Any London bird-watcher who went to Barn Elms to-day and saw nothing but gulls, tufted ducks, great crested grebes and coot would think himself ill-rewarded indeed for his pains, and Cornish's obvious delight in what he saw is a measure of the degree to which the London area has been colonised by waterfowl in the past forty years. Barn Elms has become one of the *loci classici* of British bird-watching. The author, in a comparatively few visits, has seen such remarkable sights as a grey phalarope, buoyant as a cork, bobbing on the diminutive waves of the reservoir ; a loon uttering its weird haunting wail before launching its cigar-shaped form into the air ; a vast pack of nearly two thousand pochard and tufted duck huddled together in the cold spell of 1938-39 ; and the incomparable sight of mute swans, the Catalinas of the bird world, taking off with laboured strokes of their great wings across half the reservoir, flying with ponderous grace and that matchless avian sound " hompa, hompa, hompa, hompa," over Ranelagh and Barnes, to descend again on the far side of the water with a loud swishing sound as their breasts create a sizable bow-wave.

There is no doubt that the reservoirs have brought many thousands of waterfowl to the London area that would never have come to the small amenity lakes, such as those in Osterley, Wanstead and Wimbledon Parks, or even to the flooded gravel-pits. In the census of waterfowl that was taken in the London area on December 18, 1937, which covered all the principal M.W.B. reservoirs, as well as the canal reservoirs and a good many gravel-pits and amenity lakes, the following numbers of ducks (except mallard), great crested grebes and coot were counted :

Sheld-Duck - - - - -	2
Teal - - - - - -	853
Wigeon - - - - -	272
Shoveler - - - - -	55
Common Pochard - - - -	744
Tufted Duck - - - - -	2489
Scaup - - - - - -	3
Goldeneye - - - - -	30
Long-tailed Duck - - - -	2
Goosander - - - - -	234
Smew - - - - - -	51
Great Crested Grebe - - -	385
Coot - - - - - -	1449

In his *Birds of Middlesex*, published in 1866, Harting remarks of the tufted duck that " as many as thirty have been seen at one time during hard weather upon Kingsbury [i.e., Brent] Reservoir " ; of the smew that it is " a rare winter visitant, seldom coming so far inland, except in severe weather " ; and of the coot that the greatest number observed in one party inland was forty-one, while only one goosander and five great crested grebes had ever been recorded in Middlesex at that date, none of the latter species being in breeding dress.

It is sometimes maintained that the increase of waterfowl observed in the London area is really only a reflection of the increase in observers. While there is a substratum of truth in this contention, there can be no doubt at all that there has been a very great increase in the numbers of all waterfowl that visit the London reservoirs within the last thirty or forty years. It must be remembered that many of the reservoirs, especially the larger ones, such as Staines, Barn Elms, Littleton and Chingford, are of comparatively recent construction, and

Plate XXI

Feeding black-headed gulls on Victoria Embankment with old Waterloo Bridge in the background

Plate XXII

Muswell Hill Golf Course with Alexandra Palace in the background   JOHN MARKHAM

it naturally takes time for birds to develop the regular habit of visiting them. Of all the birds which have developed this habit perhaps the most remarkable is the long-tailed duck, normally a sea-going bird, which turned up at Staines Reservoir in small numbers in each winter from 1932-33 to 1939-40, since when observation has not been possible. Hardly less striking has been the regular appearance of the goosander and the smew in some numbers.

Large numbers of gulls still, as in Cornish's day, fly out every evening to roost on the reservoirs. It is a notable sight to stand on the causeway at Staines Reservoirs at dusk, and watch the countless parties of gulls flying in to join the dark mass already roosting on the water. Long skeins can be seen far away to the east, mainly of black-heads, but also a good many of the larger types, herring, great and lesser black-backed.

At the time of the autumn and spring migrations a good many interesting waders also pass along the shores of the reservoirs, though there is rarely any inviting mud to make them stay. In 1938, for instance, ringed, golden and grey plovers, ruffs, sanderlings, knots, dunlin, a little stint, common and green sandpipers, a spotted red-shank, greenshanks, curlews and a whimbrel were all seen at one or other of the M.W.B. reservoirs. If 1937 and 1939 are also taken into account, the oystercatcher, turnstone, black-tailed godwit and purple sandpiper can be added to the list. In fact, in these three years the only waders which might reasonably have been expected to turn up at one of these reservoirs and were not in fact observed were the bar-tailed godwit, grey phalarope, curlew-sandpiper and wood-sandpiper ; all of which were recorded there at some time within the past twenty years. The only British waders, resident or regular visitors at some season, which have not occurred at one of the M.W.B. reservoirs in the past twenty years (excluding the snipes, woodcock and stone-curlew which one would not expect), are the very rare red-necked phalarope, Temminck's stint, Kentish plover, dotterel and avocet.[1]

The reservoirs seem to be fairly well stocked with fish, hence the large numbers of grebes and saw-bills, and of course the algæ and numerous freshwater animals described above provide a plentiful food supply for the other waterfowl. As the sides of the reservoirs are all bare concrete, they hold no aquatic plants and therefore none of the

[1] *Phalaropus lobatus, Calidris temminckii, Leucopolius alexandrinus, Eudromias morinellus, Recurvirostra avosetta.*

birds and insects associated with the margins of the canal reservoirs and gravel-pits.

References to Chapter 12

Bucknill (1900), Christy (1890), Cornish (1902), Glegg (1935), Harting (1866), Hastings (1937), Homes (1938), Johnson (1930), London Natural History Society (1937-44), Macpherson (1928), Ritchie (1920).

THE INFLUENCE OF REFUSE DISPOSAL

WE HAVE seen how the steady pollution of the Thames by sewage converted a fine salmon river into a waterway that stank so much as to make it almost impossible to take tea on the terrace of the House of Commons on a hot summer's day, as is now the pleasant custom. Not only the Thames, but the Fleet, Walbrook and other London streams were filled with rotting matter of the most revolting kind. Stagnant pools, like those where Gilbert White saw the martins hawking, lay in the fields all round the metropolis. Farmers from the surrounding country used to send into the City to fetch the nightsoil for manure, but it was not until after the reform of local government in the nineteenth century, and especially the creation of the Metropolitan Board of Works in 1855, that there was a proper organised public system for the disposal of refuse.

To-day London has a most efficient system of sewage disposal, and solid refuse is removed by road to vast dumps far down the estuary. The partial cleansing of the river has led to its partial re-colonisation by fish and other aquatic creatures, while the numerous sewage farms in the outer suburbs form artificial marshes where wading birds regularly resort on migration. In place of the animal scavengers of medieval London, pigs, dogs, ravens and kites, we now have rats and gulls, the latter most welcome, the former quite the reverse.

Unfortunately, not much attention has been paid recently to the fishes of the Thames. Forty odd years ago, however, C. J. Cornish was able to record a welcome return of fish from 1900 onwards. In December, 1901, for instance, a four-pound grilse was caught in the Thames estuary ; whitebait, small crabs and jellyfish reappeared at Gravesend in 1892, and three years later shoals of whitebait were seen as high up as Greenwich. Thames flounders reappeared at their favourite haunt near Chiswick Eyot, and many eels were taken between Hammersmith and Kew. In August, 1900, smelts were seen again at Putney, after an absence of many years, and in the following month a shoal passed right up the river as far as the tidal limit at Teddington Lock (Plate 20b), where many were caught. At the same time the fresh-water fish began to spread downstream, large shoals of dace, bleak, roach and small fry penetrating as far east as Putney. In 1895 roach and dace were caught below London Bridge, and in 1900 some roach reached Woolwich, though these may have come down the Lea. Cornish described scenes that cannot have been seen on the Thames in London for many years before he wrote :

"A few years ago hardly any fish were to be seen below Kew during the summer, and these were sickly and diseased. Last year (1901) they were in fine condition, and dace eagerly took the fly even on the lower reaches. Every flood-tide hundreds of ' rises ' of dace, bleak and roach were seen as the tide began to flow, or rather as the sea-water below pushed the land-water before it up the river. At high water little creeks, draw-docks, and boat-landings were crowded with healthy, hungry fish, and old riverside anglers, whose rods had been put away for years, caught them by dozens with the fly. Sixty dozen dace were taken, mainly with the fly, in a single creek, which for some years had produced little in the way of living creatures but waterside rats. I counted twenty-two ' rises ' in a minute in a length of twenty yards inside the eyot at Chiswick. During one high tide in July a sight commonly seen in a summer flood on the Isis or Cherwell was witnessed not sixty yards from the boundary stone of the county of London. The tide rose so far as to fringe several lawns by the river with a yard or two of shallow water, and the fish at once left the river and crowded into this shallow overflow, their backs occasionally showing above it, to escape the muddy clouds in the tidal water. There were hundreds of fish in the shoals, of all kinds and sizes, from dace nine inches long, with a few roach, to sticklebacks."

The purification of the river that permitted the fish to return was largely due to the efforts of the London County Council's Main Drainage Committee, which instead of allowing all the sewage of London to fall into the river at Barking and Crossness, separated the solid matter and carried it out to sea.

Not only the fish, but the birds came back to the Thames as a result of the Committee's activities, together with the preservation of the river's amenities by the Thames Conservancy. Cornish noted that the herons from Richmond Park were extending their nightly fishing excursions, which previously ended at Kew Bridge, to Hammersmith four miles farther downstream, and even to Chelsea. The scarcity of ducks on the river at Hammersmith fifty years ago can be judged from the following remark :

" In the evening and early morning a few wild ducks accompany the herons as low as the reach above Hammersmith Bridge, and single ducks have been seen even at midday flying overhead."

To-day ducks are nowhere more abundant on the Thames than on the short stretch between Hammersmith Bridge and Chiswick Eyot. (Plate XX.) Mallard, pochard and tufted duck, sometimes totalling hundreds, can be seen on almost any visit in the winter months, along with a great concourse of a hundred or more swans. (See Plate 20a.) The ducks are constantly flying to and fro between the river and the adjacent Barn Elms and Lonsdale Road Reservoirs.

Since the war has restricted the visits of London bird-watchers to the large reservoirs, the Thames towpath between Hammersmith and Richmond has become one of their favourite resorts, where in most winters ten or a dozen different species of duck are seen. In 1943, for instance, the list included mallard, gadwall, teal, garganey, wigeon, pintail, pochard, tufted duck, scaup, goosander and smew, and in 1942 the sheld-duck, shoveler, goldeneye and red-breasted merganser were seen. Grebes and divers are of less frequent occurrence, but cormorants and shags appear in most years. During the cold spell in February, 1940, two brent geese (*Branta bernicla*) were seen on both sides of the river near Chiswick Eyot, but these may have been escapes.

Even right in the heart of London, between Westminster and Blackfriars Bridges, mallard occur all the year round, and a pair once nested on a decayed raft near Lambeth Bridge. In the winter tufted ducks come to the river near Waterloo and Chelsea Bridges, and are fed by the crowds on the Embankment. Mr. Johnson has recorded several fishes and other aquatic creatures from the Thames at Battersea within recent years, including a miller's thumb (*Cottus gobio*) in 1911, a jack and two or three small perches in 1927, and barnacles on the timber of Battersea Wharf. London's river is undoubtedly recovering its attraction for the naturalist even in the heart of the built-up area.

Instead of going direct into the rivers and streams, the sewage of the outer districts of London now passes through a multitude of sewage farms, large and small, such as those at Beddington, Epsom and Brooklands in Surrey, Elmers End in Kent, Romford and Chigwell in Essex, Watford in Herts, and Harrow and Edmonton in Middlesex. Here the sewage is filtered through many shallow pools, which form ideal feeding-grounds for waders, while the relatively secluded surroundings provide breeding places for other water-loving birds, such as yellow wagtails, reed-buntings, sedge-warblers and moorhens. So well known are sewage farms among bird enthusiasts as the haunts of rare birds, that it is on record that a lady once asked the manager of a sewage farm near London, which she was visiting with a party of naturalists, what sewage farms were really for.

During the years immediately before the war the assiduity of London bird-watchers recorded many species of waders from the sewage farms in the London area. In 1938 alone, for instance, thirteen species, including the ringed and golden plovers, ruff, dunlin, common and green sandpipers, greenshank, curlew, whimbrel and jack snipe were seen, and if 1937 and 1939 are taken into account, the wood-sandpiper, spotted redshank, black-tailed godwit, curlew-sandpiper and little stint can be added to the list. At Brooklands sewage farm, which lies within the famous racing track, such rarities as the avocet (*Recurvirostra avosetta*) and Temminck's stint (*Calidris temminckii*) have been seen.

Huge numbers of birds come to the sewage farms to feed every day in the autumn and winter. Thousands of gulls, starlings and finches feed on the filter-beds and on the dense growth of goosefoot (*Chenopodium*), persicaria (*Polygonum persicaria*) and other plants that grow in the disused filter-beds. At the extensive sewage farm of the Croydon Corporation at Beddington, gulls come daily from the reservoirs at Staines and Littleton, ten or fifteen miles away, while starlings fly in from all sides, some of them also roosting more than ten miles away.

The process of sewage purification creates a unique kind of habitat for insects on a scale hitherto unsurpassed in the world. When crude sewage arrives at the farm or other type of works, the grosser solid matter is removed, and the resultant liquid passed through a series of sprinklers, spreading over beds of stones which contain bacteria.

Incidentally, starlings have a particular liking for these sprinkler beds, and may be seen in large numbers perched on the slowly revolving sprinkler, while others fussing about below dodge smartly through the curtains of spray which bear down on them. In the beds of stones the liquid sewage gradually becomes purified, largely as a result of oxygen absorption, and the effluent can then safely be passed out into a river or stream. On some of the older farms, most beloved by wading birds, the liquid also seeps through fields, which thus become a cross between the Hampshire water meadows and undrained marshes.

The plentiful moisture and plant food on the bacteria beds results in the development of a thick rind of plant growth, largely made up of the blue-green alga *Phormidium*. This in turn provides abundant food material for the larvæ of several kinds of flies and worms; in fact if for any reason there are no flies or worms the algæ grow so strongly as to choke the beds. The insects chiefly involved in this unsavoury way of life are three chironomid midges (*Metriocnemus longitarsus*, *M. hirticollis* and *Spaniotoma minima*) and two psychodid flies (*Psychoda alternata* and *P. severini*); an enchytræid worm (*Lumbricillus lineatus*) is also common. Other denizens of sewage beds include the fly *Anisopus fenestralis*, which is prone, as its name suggests, to infest windows in houses near sewage farms in some numbers, and the springtail *Achorutus viaticus*.

Mr. L. Lloyd has given a vivid picture of the zonal distribution of the fauna of filter-beds :

". . . each piece of clinker was capped by a heap of slimy debris which was a wriggling mass of enchytræid and nematode worms, while a few inches down in the medium was an almost continuous layer of developing Psychoda and Anisopus flies."

Household and general rubbish from London is mostly taken down the Thames in barges and dumped on the marshes between Barking and Tilbury; smaller quantities are tipped at Yiewsley on the Grand Union Canal and on Hackney Marsh adjacent to the Lea. In addition to providing an abundant breeding ground for thousands of rats and crickets, these rubbish dumps are a fruitful hunting ground for botanists, especially those in search of adventive plants. About twenty years ago five dumps in the London area, Dagenham, Grays and Tilbury in Essex, and Hackney Marsh and Yiewsley in Middlesex were studied from this point of view by Messrs. R. Melville and R. L. Smith, who found altogether some 250 adventive plants growing on them. The Dagenham area was the largest, covering a square mile on the Thames bank between Dagenham Dock and Rainham, and containing in addition to an abundant native flora, about 170 alien adventives. Two of the aliens, the giant hogweed (*Heracleum mantegaz-zianum*) and a dock *Rumex patientia*, grew together and formed " a veritable forest of vegetation over eight feet high that must be seen to be appreciated." Large areas of the newer parts of the dump were covered with a dense undergrowth of black nightshade (*Solanum nigrum*) and various goosefoots, especially *Chenopodium rubrum*. At Yiewsley a large variety of cereals were found, probably mainly derived from chicken-food refuse. Other interesting aliens included several common garden plants like the delphinium (*Delphinium gayanum*), Californian poppy (*Eschscholzia californica*), night-scented stock (*Matthiola bicornis*), mignonette (*Reseda odorata*) and hollyhock (*Althæa rosea*), some food refuse plants such as the grape vine (*Vitis vinifera*), common gourd (*Lagenaria vulgaris*), water melon (*Citrullus vulgaris*), caraway (*Carum carvi*) and tomato (*Solanum lycopersicum*), and such escapes from the kitchen garden as the globe and Jerusalem artichokes (*Cynara cardun-culus, Helianthus tuberosus*), endive (*Cichorium endivia*), lettuce (*Lactuca sativa*), beetroot (*Beta vulgaris*), and spinach (*Spinacea oleracea*). A native of India and Africa that was found at both Dagenham and Hackney Marsh was *Guizotia abyssinica* ; this is cultivated in India for its seeds, which yield a bland oil similar to sesame.

Rats, which abound in rubbish dumps, are also abundant in the sewers of London, where some 650,000 brown rats (Plate VIa) were killed in an intensive anti-rat campaign in the winter of 1943-44.

London has never been without its bird scavengers, except for the short interval between the loss of the kites and ravens and the appearance of the gulls, a period of a little more than a hundred years. It was as scavengers that the gulls first came to London, though of late years they have far transcended their original feeding habits, and are now regular visitors to the golf-courses and playing fields, as well as to the river, sewage farms and refuse dumps.

The rise of the gull family to be typical urban birds is one of the most striking developments in the history of British ornithology, and it has all happened within the past fifty years. When Harting wrote in 1866, only the black-headed gull was at all regularly observed in

Plate XXIII

A typical suburban playing field   JOHN MARKHAM

Plate XXIV

The melanic form of the peppered moth S. BEAUFOY

A London stray cat JOHN MARKHAM

Middlesex, and then it was only on passage in spring or autumn or after a gale that they were found so far inland. It seems probable, however, that even at this time gulls of various sorts were regular visitors to the London docks in small numbers ; unfortunately all too few London ornithologists have paid attention to the docks.

At all events, a succession of severe winters from 1887-88 to 1894-95 led to large numbers of gulls, chiefly black-headed, coming up the Thames as far as Putney. From the latter year onwards, black-headed gulls have been regular visitors to the Inner London stretches of the Thames, and have gradually extended their range away from the river until there is hardly any part of the Greater London area to-day where these gulls may not be seen in the winter months. Four other species of gull have followed the black-head, which incidentally belies its name for most of the period when Londoners see it, and are now regular winter visitors or passage migrants in the London area. These are the common, herring, lesser black-backed (both British and Scandinavian sub-species) and great black-backed gulls. The great black-backs are rarely seen away from the river and larger reservoirs, but the others can often be seen on playing fields and sewage farms.

W. H. Hudson has left us a graphic picture of the black-headed gulls in St. James's Park :

" It was a rough wild morning ; the hurrying masses of dark cloud cast a gloom below that was like twilight ; and though there was no mist the trees and buildings surrounding the park appeared vague and distant. The water, too, looked strange in its intense blackness, which was not hidden by the silver-grey light on the surface, for the surface was everywhere rent and broken by the wind, showing the blackness beneath. Some of the gulls—about 150 I thought—were on the water together in a close flock, tailing off to a point, all with their red beaks pointing one way to the gale. Seeing them thus, sitting high as their manner is, tossed up and down with the tumbling water, yet every bird keeping his place in the company, their whiteness and buoyancy in that dark setting was quite wonderful. It was a picture of black winter and beautiful wild bird life which would have had a rare attraction even in the desert places of the earth ; in London it could not be witnessed without feelings of surprise and gratitude."

The gulls came to London as scavengers, picking up what they could from the solid matter floating down the Thames. They soon, however, adapted themselves to other habitats. Large numbers found sustenance in the scraps offered them by a benevolent public along the Embankment and in the parks. (Plates 21a, XXI.) Others found an important source of food at the sewage farms and rubbish dumps of the London area. Yet others began to forage on the many golf-courses and playing fields (Plates XXII, XXIII) around London. As scavengers still they have found their way into many railway sidings and marshalling yards. In this habitat they have, along with the house-sparrows and London pigeons, developed into completely urban birds, entirely dependent on human activities for their food.

The railway gulls, along with the great majority of the others in the London area, fly out to another man-made habitat, the great reservoirs in the Thames and Lea valleys, to roost in the evenings. It is hard to conceive of an avian way of life more fundamentally influenced by man than that of a black-headed gull which spends its daylight hours at Willesden Junction and its nights roosting on Staines Reservoir. It is probable that there are, in fact, gulls which confine themselves to scavenging on the railways, though this remains to be proved by some enterprising platelayer-ornithologist. Mr. T. L. Bartlett has shown, by catching gulls, ringing them and catching the same ones again, that gulls which are caught on the river are always recaptured in the same place on the river, while those ringed in the parks are caught again only in the parks. The life of a bird is a good deal narrower and more stereotyped than might be imagined from the freedom of movement that its wings give it.

References for Chapter 13

Bartlett (1944), Cornish (1902), Harting (1866), Henson (1944), Hudson (1898), Johnson (1930), Lloyd (1943), London Natural History Society (1937-44), Melville & Smith (1928), Rowberry (1934), Stubbs (1917).

THE INFLUENCE OF SMOKE

" I love the very smoke of London, because it has been the medium most familiar to my vision."—CHARLES LAMB.

———

WE saw in Chapter 5 that complaints about the ill-effects of the London smoke have been loud for the past three hundred years. It is only to-day, however, that we are beginning to realise the full significance, in quantitative terms, of the pollution of London and its countryside by smoke.

In Central London as much as 322 tons of solid matter per square mile was deposited in 1936-37 at Archbishop's Park, Lambeth. This total included forty-three tons of sulphates and five tons of tar. Even as far out as Finsbury Park in the north-east and Ravenscourt Park in the west, both on the borders of the County of London, the annual deposit of solid matter in the same year amounted to 273 tons and 210 tons respectively. If it is assumed that there was a uniform deposit of 250 tons over the whole county, then in a single year nearly 30,000 tons of solid matter would be deposited from the smoky atmosphere on the County of London.

Another measure of the degree of pollution of the atmosphere is the concentration of sooty suspended matter. In the winter months of 1936-37 this concentration amounted (in milligrams per 100 cubic metres of air) to 69 in Victoria Street and 65 in Charing Cross Road, compared with 31 in South Kensington and 15 at Kew.

One of the most important consequences of this pollution is the prevention of sunlight from reaching the ground in Central London. The following table shows how much sunshine Central London (Bunhill Row off City Road) loses compared with Kew and Greenwich in the suburbs :

MEAN DAILY BRIGHT SUNSHINE IN HOURS

Station				1936	1937	1938	Normal
Bunhill Row	-	-	-	2·97	3·07	3·54	3·29
Greenwich	-	-	-	3·20	3·13	3·68	4·01
Kew	-	-	-	3·57	3·72	4·00	4·01

The serious effect of atmospheric pollution on human health has long been known. Evelyn complained that on account of the excessive smokiness of the London air,—

". . . almost one half of them that perish in London dye of Phthisical and Pulmonic distempers ; . . . the inhabitants are never free from Coughs and importunate Rheumatisms. . . ."

Modern statistics confirm Evelyn's empirical judgment. In 1934, for instance, when there was fog from November 10 to December 1 in London, deaths from respiratory diseases rose from forty-nine in the week before the fog to 121 in the last week of the foggy period.

Animals also suffer. During a foggy period in December, 1873, some of the prize beasts at the Islington Cattle Show were suffocated, and others had to be slaughtered to put them out of their misery. According to Colonel A. E. Hamerton, Pathologist to the Zoological Society of London, much damage is done to the health of the animals in the Zoo in Regent's Park by the polluted air :

" Animals in confined spaces suffer most from dust, especially in foggy weather. The pathological changes known as anthracosis, caused by deposition of carbon particles, and silicosis, caused by deposition of siliceous particles, in the pulmonary tissues and their associated lymph glands, progress gradually and end fatally after some years of life in a dusty atmosphere.

" Many felines in the lion house die from chronic bronchitis, fibrosis and gangrene of the lungs associated with blackening of the lungs by dust deposit. Among birds that have lived for some years in the aviaries, fatal necrosis of the lung localised around carbon deposits causes many deaths ; such necrotic foci in the lung are frequently colonised by tubercle bacilli which rapidly become disseminated with fatal results."

ERIC HOSKING

Almond blossom in suburban front gardens, Ruislip, Middlesex

W. SUSCHITZKY

Roses in Queen Mary's Garden, Regent's Park

PLATE 32

Pear tree in blossom, Crouch End, Middlesex

Another serious consequence of atmospheric pollution from the health point of view is the reduction in the amount of sunshine enjoyed by those who live in smoky areas, which involves among other things a reduced manufacture of vitamin D by the body.

Even worse than its effect on animals is the influence of a smoky atmosphere on plant life. The effects of smoke on vegetation have been summed up by Dr. A. G. Ruston under four main heads :

(1) The cloud of smoke blocks out the sunlight, and reduces the available solar energy, which is needed by green leaves to convert carbon dioxide into carbohydrates, by anything up to 40 per cent ; this results in stunting the growth of any vegetation that survives in the area.

(2) The tarry matter that coats over the leaves chokes the stomatal openings and so checks or even prevents the natural process of transpiration and assimilation ; conifers and other evergreens are especially affected, and in some cases, such as the privet, will become deciduous before dying ; plants whose leaves have a crinkled or hairy surface which catches the soot (e.g., calceolaria, primrose, hollyhock) always do badly in a sooty atmosphere, but those whose leaves are hard, smooth or leathery, like pinks, carnations, irises and London pride (*Saxifraga umbrosa*) are highly resistant, which doubtless explains how the last-named plant came to acquire its name.

(3) The presence of free acids in the air tends to lower the reproductive capacity of plants and to deprive them of the power to produce colour ; for instance bronze flowers in a smoky district gradually turn yellow.

(4) The free acids in the soil make it sour, and thus limit the activity of the soil organisms, which must work freely if the soil is to maintain its fertility.

The fogs and smoke of London have a markedly harmful effect on grass, causing a slimy scum and creating acid conditions that destroy the useful bacteria of the soil, which in turn results in a coarse growth of grass ; moreover, when the grass is rolled, as on a tennis court, weeds tend to grow rapidly. The deposit of soot has had such a grave effect on the conifers in Kew Gardens that the authorities have had to set aside an area of land near Tunbridge Wells, Kent, for the formation of a National Collection of Conifers. The late Sir Arthur Hill, Director of the Royal Botanical Gardens, has recorded that a single bad fog at Kew caused practically all the flowers and unopened buds of orchids to fall in less than twenty-four hours, while the leaves of begonias and other plants also fell in large numbers. He also stated that after a bad fog it costs about £100 to wash the glasshouses, so that the full benefit of the sunlight can be enjoyed by the plants again. Dr. Bewley, Director of the Experimental and Research Station at Cheshunt in the Lea valley, finds that the glass of the glasshouses in Ponders End and Enfield districts is continually being blackened by smoke and has to be washed over every spring. A heavy fog can leave such a deposit of soot as entirely to destroy the crop of young tomato seedlings. Many nurserymen have been compelled to move farther up the Lea valley away from London in order to reach purer air (Plate 40a).

Atmospheric pollution is probably one of the factors responsible for the scarcity of many kinds of wild plants near London. The relative infrequency of lichens on the trees in Epping Forest, for instance, has been attributed to this cause.

The pollution of the air by smoke and other noxious vapours is due to preventable causes, entirely under human control, the most important of which is the burning of soft bituminous coal, especially in open domestic grates. There is no more direct biotic influence of man among all the many discussed in this book, and no more easily remediable one, than atmospheric pollution.

An interesting indirect effect of the smokiness of London's atmosphere is the so-called industrial melanism in moths. A number of moths, notable the waved umber (*Hemerophila abruptaria*), mottled beauty (*Boarmia repandata*), grey dagger (*Acronycta psi*) and peppered moth (*Pachys betularia*) (Plate XXIVa), produce black or blackish variants in London and other industrial districts in England. Following Ford, the cause in now generally held to be due to a factor of natural selection. Melanic forms, where dominant (as is the case with all industrial melanics), are rather hardier than ordinary ones, but in most areas this advantage is more than cancelled out by their greater conspicuousness, so that predators destroy them more quickly than the normal forms. In sooty areas, such as London, however, their darkness gives the melanic forms an advantage, for they are now less conspicuous than the lighter normal types, and this together with their natural hardiness has led to a considerable increase, sometimes amounting to

100 per cent replacement, of melanic types during the past fifty years or so.

One of the chief factors in reducing the amount of smoke pollution in recent years has been the widespread substitution of gas and electricity for coal in domestic use. The large power stations and gasworks that produce the electricity and gas have another quite different biotic influence insofar as they provide oases for wild life in heavily built-up areas. Kestrels, for instance, have often been suspected of nesting on gas-holders, though this is hard to prove owing to the height of the gas-holders. Mr. Johnson records that a kestrel's nest was found by steeplejacks on one of the four chimney-shafts of the London County Council's tramway power-station at Greenwich in 1928.

A typical " gasworks sanctuary " is that surrounding the large gas-works at Bromley-by-Bow at the mouth of the Lea (Plate XXV). It consists of a quarter of a square mile of relatively open land, lying in the heart of the industrial East End, adjacent to Bow Creek. In addition to the nine gas-holders and various buildings, there are some allotments, playing fields, a children's garden and playground, a grove of poplars, some sycamores and several fruit-trees, a good deal of waste ground, and at the south end some waste lime dumps rising to a height of about 30 feet and then sloping down again to a small semi-tidal marsh, where rushes grow.

In this unpromising locality Mr. R. P. Donnelly of the Gas Light and Coke Company has recorded (1938) a most interesting series of animals and plants. Elder bushes (*Sambucus nigra*) grow freely, and there are one or two small wayfaring-trees (*Viburnum lantana*). Among the flowering plants may be mentioned the scarce danewort or dwarf elder (*Sambucus ebulus*), the ubiquitous rose-bay willow-herb (*Epilobium angustifolium*) and coltsfoot (*Tussilago farfara*), an abundance of the stinking mayweed (*Anthemis cotula*), and others, such as penny-cress (*Thlaspi arvense*), alexanders (*Smyrnium olusatrum*), evidently a relic of the former semi-maritime Thames-side flora, goutweed (*Aegopodium podagraria*), shepherd's needle (*Scandix pecten-veneris*), yarrow (*Achillea millefolium*), chicory or succory (*Cichorium intybus*), and bittersweet or woody nightshade (*Solanum dulcamara*).

The mammals of the area include hedgehogs and a colony of rabbits. The rabbits appear to be lighter in colour than ordinary wild ones, and may be a survival from the time when the whole area was open country. The hedgehogs come at night to be fed with milk.

The surprisingly large list of birds includes skylarks, stonechats, woodpigeons (a pair nested in a stunted tree on the lime dumps in 1936), pied wagtails, goldfinches, kestrels, carrion-crows (which nest on the gas-holders), moorhens, starlings (which roosted in the poplars in some thousands in October, 1937), and the usual sparrows, gulls and London pigeons.

The insects recorded within the gas-works compound are few, but include an occasional red admiral (*Vanessa atalanta*), painted lady (*V. cardui*), and stag-beetle (*Lucanus cervus*). Caterpillars of the poplar hawk-moth (*Smerinthus populi*) are often found on the poplar-trees.

References for Chapter 14

Ford (1940), Johnson (1930), London Natural History Society (1900-44), Mera (1926), National Smoke Abatement Society (1938, n.d., n.d.a.), Owens (1938), Robbins (1916), Ruston (1936).

Plate XXV

Waste ground at Bromley-by-Bow gasworks JOHN MARKHAM

Plate XXVI

A typical backyard poultry run   JOHN MARKHAM

THE INFLUENCE OF FOOD-GETTING

THE WHOLE natural history of London, as traced in the preceding chapters, has been one of the steadily decreasing influence of food-getting by man as a biotic factor. At the present day there is hardly any species of wild animal or plant in or near London whose status is undergoing serious modification on account of its use for human food. Moreover the food that is artificially cultivated within twenty miles of St. Paul's would not now feed more than a minute fraction of the human population inhabiting the same area, though oddly enough it would suffice to feed the remaining inhabitants of the old City area, who in 1938 numbered only 9380.

The first regular agricultural statistics for Middlesex, which then included all that part of the County of London which lies north of the Thames, were recorded for 1866, when 61 per cent of the county's 180,136 acres were still under crops or grass, including 38,736 acres of arable. Not until 1893 did separate figures appear for the County of London, at which time some 20 per cent of the total of 75,442 acres were still cultivated, mainly in the south-eastern boroughs of Woolwich and Lewisham. The last separate figures for the County of London were published in 1922, when the area under crops and grass had declined to about 7 per cent. Nevertheless, it may surprise modern Londoners to know that only twenty-two years ago there were still 1636 acres of arable land, 3447 acres of permanent grass, and 887 acres of rough grazing within their so heavily built-up urban area.

The most recent statistics are for 1938, when the 33,105 cultivated acres of the Counties of London and Middlesex represented 15 per cent of their joint area of 219,990 acres. Of this area under crops and grass, 11,139 were arable, 5707 under permanent grass for hay, 16,259 under permanent grass for grazing, and 3577 represented rough grazings. It seems a little odd to find the category " rough grazings," reminiscent of windswept Welsh hillsides and Cornish sea-cliffs, in

London and Middlesex at all. The explanation is to be found partly
in the many commons, partly in the large amount of land scheduled
for building but not yet built on, and partly in the fact that farmers are
in the habit of balancing the nominal and returned acreages of their
farms by putting down so much rough grazing to make the sum add
up correctly. The very careful findings of the Land Utilisation Survey
for London and Middlesex suggest that the definition of " agricultural
holding " for the purpose of the agricultural returns leads to some
inaccuracy in representing the true facts of the use of the land in these
very urban counties. The following table shows the difference between
the two sets of figures for the years 1932-35 :

Type of Land Use	Ministry of Agriculture	Land Utilisation Survey
	Percentage of area of London and Middlesex	
Forest and woodland - - -	0·9 (1924)	1·5
Arable - - - - - -	6·8	5·2
Permanent grass - - - -	13·2	26·4
Rough grazing - - - -	2·0	1·2
Orchards - - - - -	0·9	1·2
Houses with gardens - - -	Not shown	39·4
Nurseries - - - - -	,,	0·9
Agriculturally unproductive - -	,,	24·2
Unaccounted for - - - -	76·2	0

They differ chiefly in the much greater amount of permanent grass
shown by the Land Utilisation Survey ; this was probably due to the
inclusion of a good deal of grassland normally used only for amenity
purposes, such as playing fields and golf courses, in which the London
area is particularly rich.

The agricultural returns for June 4, 1938, show in considerable
detail the kind of farming still carried on in London and Middlesex,
all of it within the twenty-mile radius of St. Paul's. The principal crops
were potatoes (1082 acres), wheat (883 acres), oats (589 acres), beans
(407 acres) and peas (342 acres). Small quantities of rye, turnips,
swedes, mangolds, vetches, tares, barley, mixed corn, lucerne, cabbages
for fodder, and clover and rotation grasses for hay were also grown.
Some 10 per cent of the whole arable acreage was fallow, which may
be compared with the 33 or even 50 per cent that would have been

found in the same area in the year 938. The flat south-western part of Middlesex is still a great market-gardening area, and many vegetables besides peas and potatoes were grown, such as 1437 acres of cabbages and other greens, 561 acres of lettuces, 254 acres of rhubarb, 152 acres of cauliflowers and broccoli, and smaller amounts of carrots, onions, celery and asparagus.

The orchards of Middlesex, of which nearly 1000 acres survived in 1938, grow mainly apples, cherries and plums. (Plate 23b.) Some 200 acres of small fruit in the same year included 87 of gooseberries, 46 of raspberries, 34 of strawberries, 28 of red and black currants, and 9 of loganberries and cultivated blackberries. Of 360 acres devoted to the growing of flowers, 54 went to daffodils and narcissi and 26 to other bulb flowers. Of 220 acres of glasshouses, chiefly in the Lea valley, 81 were devoted to tomatoes.

Livestock farming is still carried on in London and Middlesex (Plate 40a) and the following table gives the number of livestock on agricultural holdings (i.e., excluding backyard poultry) (Plate XXVI) in 1938 :

Horses	-	-	-	-	-	1,762
Cattle	-	-	-	-	-	7,746
Sheep	-	-	-	-	-	5,799
Pigs -	-	-	-	-	-	19,116
Fowls	-	-	-	-	-	57,019
Ducks	-	-	-	-	-	1,622
Geese	-	-	-	-	-	834
Turkeys	-	-	-	-	-	350

In London domesticated livestock, especially since the decline of horse transport, do not play a very large part in the animal community. A crowd of sparrows feeding on horse-droppings is no longer the everyday sight it was when the clip-clop of a hansom in foggy Baker Street was soon followed by the ring at the bell of No. 221b, and Holmes's inevitable " And there, if I am not mistaken, Watson, is our visitor." As late as 1905, long after *A Scandal in Bohemia*, cows still appeared daily in St. James's Park to supply their milk direct to the consumer. Cows in byres have survived longer ; in 1889 there were 10,000 cows in 745 byres in the County of London, but to-day only a handful survive in the East End, though on the testimony of Mr. Johnson there were ten cows in a small byre in Clipstone Street, off Euston Road, as late as 1926. There is still a haysel in Green Park (Plate XXVII), and flocks of sheep are still grazed in Hyde Park and on

Hampstead Heath (Plate 22), and before the war were brought annually from Scotland by train for this purpose. It is only in wartime that pigs and poultry have reappeared in Inner London (Plate 13b), and the Welch Regiment and the Zoo authorities are the only people on record as having kept goats there in the inter-war years.

All round London, in the suburban fringe, agriculture manages to hang on until the building price of the land sufficiently outweighs the relatively meagre farming profits. Dairy farms existed at Dulwich and Highgate, both within five miles of Charing Cross, till after 1918. One of the very last real farms in the County of London is shown in Plate 23a ; it is Coldharbour Farm at Mottingham, on the road to Chislehurst.

This is not the place to discuss the animal and plant communities associated with agricultural land, in itself a fascinating subject and containing material for a full volume. It is enough here to refer to the many bird and insect pests of farm and garden produce, to the rich and interesting weed flora of cornfields and market gardens, and to the effect of grazing cattle on the vegetation of pastures. None of these phenomena are found to any specially notable extent in the London area. Woodpigeons and hawfinches, as elsewhere, pillage the peas ; blackbirds and thrushes, as elsewhere, take toll of the cherries and soft fruit, and starlings also have been recorded as destroying pears, apples and plums at Eltham, and devastating raspberry fields at Swanley, Kent ; wasps guzzle at the plums ; wireworms (at the rate of 600,000 per acre in Middlesex) chew up the plant roots ; the frit-fly infests oats.

The goldfinch (Plate 24a) is an interesting example of a bird that is closely associated with the food-getting activities of man. In the London area it is rarely found nesting far from human habitations, and is especially fond of orchards in the breeding season. Hence it is especially common in the fruit-growing districts of North-west Kent and South-west Middlesex. In the Farnborough district goldfinches are most often found nesting in one particular kind of plum-tree, and only rarely in apple-trees. Elsewhere, goldfinches nest chiefly in gardens, hedgerows, oak copses and open commons, and individual nests have been recorded in such diverse situations as pine-trees, a high beech-hedge, elder and laurel bushes and fifteen feet up in a horse-chestnut.

Outside the breeding season goldfinches are most often found either in alder-trees or on waste ground where seeding thistles and other composite plants grow. Out of fifty-six separate feeding records for goldfinches in the London area, eighteen refer to thistles of various kinds, sixteen to alders, twelve to other named plants, and the remainder to birch-trees, conifers, insects and fields. In addition to four kinds of thistle (*Silybum marianum, Cirsium lanceolatum, C. palustrc, C. arvense*), composites such as the dandelion (*Taraxacum officinale*), groundsel (*Senecio vulgaris*), ragwort (*S. jacobæa*), great and black knapweeds (*Centaurea scabiosa, C. nigra*), cornflower (*C. cyanus*) and golden-rod (*Solidago*) are favoured by feeding goldfinches in the London area, while the teasel (*Dipsacus sylvestris*), angelica (*A. sylvestris*), chickweed (*Stellaria media*), persicaria (*Polygonum persicaria*), greater plantain (*Plantago major*) and shepherd's purse (*Capsella bursa-pastoris*) have also been recorded.

The goldfinch thus supplies an instructive example of a bird that has partly, but by no means wholly, adapted itself to the changes of environment brought about by cultivation. Insofar as it still feeds on alders and birches, it is clinging to the sort of environment it would have found in the lower Thames valley two thousand years ago, when there must have been as many alders within ten miles of Charing Cross as there are privet bushes now. It seems unlikely that all the goldfinches that feed on alders and birches in the London area in the winter can be immigrants from other parts of the British Isles—the Continental race has not yet been proved to visit Britain—and in any case even away from London goldfinches have adapted themselves pretty thoroughly to a cultivated habitat in the breeding season.

Two minor cultivating industries in the London area are worth a mention, the growing of osiers and watercress. When Cornish wrote forty years ago, osiers were still cultivated for sale on Chiswick Eyot (Plate XX), and on the testimony of Messrs. G W. Scott and Sons, basket-makers of Charing Cross Road, willow " sticks," used for the staking or ribs of a basket, were obtained from this area as late as the mid-1930's. In 1902 the osier shoots were in great demand for making baskets, crates, lobster-pots and eel-traps, the harvest lasting from January to March each year :

" On Chiswick Eyot, which is entirely planted with osiers, there are standing at the time of writing six stacks of bundles set upright. Each stack contains about fifty bundles of the finest rods, nine feet high. Thus the eyot yields at least three hundred bundles. This osier-bed is cut quite early in the year, usually in January, and by February all the fresh rods are planted."

The grass that grew among the osiers used to be sold to local horse-owners, and at one time also to dairymen. Even to-day both sedge and reed-warblers breed on Chiswick Eyot, their nearest point to Charing Cross. In the summer of 1900 a young cuckoo was reared in a sedge-warbler's nest on the Eyot.

The growing of watercress is an increasingly important form of market gardening round London, especially on the chalk streams. Cress beds have been formed on the Wandle at Carshalton, on the Mole at Fetcham, and on the Misbourne at Denham, among other places within twenty miles of St. Paul's. Kingfishers, moorhens, dab-chicks and other aquatic birds haunt the streams that run through the cress-beds, and starlings may be seen fussing busily over the cress.

Turning to the use of wild animals and plants for food, we find that only rabbits and hares among the mammals are so used. Especially in wartime, rabbits are always a welcome addition to the larder, and shooting them for the pot may exercise an appreciable biotic effect.

R. W. Robbins has described an interesting chain of cause and effect, linking the 1914-18 war and the subsequent building boom with the disappearance of a colony of butterflies, that has the snaring of rabbits for food as one of its links. The colony of butterflies consisted mainly of chalkhill blues (*Agriades corydon*), with a few adonis blues (*A. bellargus*), both of whose larvæ feed on the local horseshoe vetch (*Hippocrepis comosa*), on the scarp of the North Downs near Oxted, Surrey. In 1920 Robbins found these butterflies plentifully on the floor of an old chalk-pit just to the east of the great chalk-pit that scars the Downs above Oxted. By 1926, however, both species had completely disappeared from the locality, which was their only station for several miles around. Indeed, there were no butterflies of any kind to be seen at the spot. A close inspection of the turf and herbage showed that it was close-bitten. The only plants standing above the turf were centaury and yellow-wort (*Centaurium umbellatum, Blackstonia perfoliata*), and the autumn gentian (*Gentiana amarella*), all members of the gentian family and containing a bitter principle. Though rosettes of the stemless thistle (*Cirsium acaule*) were abundant, each one had the central bud or flower bitten out. No trace of the horseshoe vetch could be found, either on the floor or on the sides of the pit. Traces of the culprits, on the other hand, could be seen every-where, in the shape of rabbits' pellets.

Rabbits have always been fairly plentiful on the Surrey hills, and though shooting them was forbidden after the neighbouring woods began to be preserved for game again after the war, Robbins con-sidered the chain of cause and effect to be more complex than this. During the war the cottagers nearby, who worked in the large chalk-pit, had been very glad to snare the rabbits to supplement their rations. This unwonted attack on their numbers kept the rabbits down so much that the horseshoe vetch was able to increase until it could support a colony of chalkhill and adonis blues. After the war, how-ever, there was a building boom, the demand for lime was brisk, and the combination of regular work and good pay for the cottagers with the end of food rationing led to a sharp easing of the pressure on the local rabbit population. Thus they increased again, and wiped out both the horseshoe vetch and the blue butterflies whose larvæ fed on it.

There is virtually no shooting of birds for the pot nowadays in the London area, though birds may be obtained for the pot both as a by-product of sport or as a result of the destruction of such farmers' banes as the woodpigeon. The custom of shooting or netting small birds for the pot seems to have died right out, though *The Story of San Michele* shows that it still prevails on the Continent. Harting, writing in 1889, said that large numbers of skylarks were netted by bird-catchers on Hampstead Heath, but the custom seems to have died out by 1913, and there is no reason to suppose that the larks which form an in-gredient in the Cheshire Cheese's famous steak and kidney pie are netted anywhere near London.

The motive of most anglers round London is also sporting rather than food-getting, though naturally a tasty perch is not spurned. As recently as forty years ago, however, as C. J. Cornish has described, there was an eel fishery on the Thames at Hammersmith and Fulham, carrying on a trade that had links with Domesday. In one night sixteen eels were netted, and a Chiswick fisherman once brought Cornish an eel weighing four pounds. Other fish, such as dace, roach and smelts, were also caught by these last of the Thames fishermen. They were said to be sold to the London Jews, who apparently preferred freshwater fish, and would pay twopence each for the dace. Even within the past few years eels have been caught in the ponds in Kew Gardens, and eaten by members of the staff.

Wild plants, especially those bearing fruits, such as blackberries and hazel-nuts, are still sought out by Londoners in the autumn. Wartime shortages of fruit accentuated the demand for black-berries, and the County Council gave permission for the picking of blackberries on Hampstead Heath and in Ken Wood. It is on record that the hazel was once almost rooted out of Epping Forest to prevent the disturbance of the deer by nutting parties from London. Other wild fruits still obtainable near London include elderberries and sloes, though few now make elderberry wine or sloe gin, and wild strawberries and raspberries, the last-named of which grow in plenty on the remaining outskirts of Bishop's Wood in Finchley, only about five miles from Charing Cross. According to Ramsbottom, five or six kinds of fungi could be obtained at Covent Garden market eighty years ago, though this compares very poorly with the fungus market in German and other European towns. To-day mushrooms have been exterminated from many fields near London by people with a taste for the only fungus that is at all widely eaten in modern Britain.

The war has brought a demand for many home-grown herbs that were formerly imported. Among the plants called for by the Direc-torate of Medical Supplies which grow commonly in or near London are the foxglove (*Digitalis purpurea*), male fern (*Dryopteris filix-mas*), stinging nettle (*Urtica dioica*), dandelion (*Taraxacum officinale*), burdock (*Arctium* spp.) and coltsfoot (*Tussilago farfara*). Other important drug-making plants are scarcer in the London area. The true valerian (*Valeriana officinalis*) has been recorded in woods on the chalk at Farthing Downs, Coulsdon, Surrey, and east of Downe, Kent. The deadly nightshade or belladonna (*Atropa belladonna*) is known at several places on the North Downs, and has been recorded as near the centre of London as North Hill, Highgate, and Chingford, Essex. Henbane (*Hyoscyamus niger*) has occurred casually at Uxbridge in Middlesex, Chingford, Hale End, Woodford Green and Grays in Essex, and Dartford in Kent.

References for Chapter 15

Bishop *et al.* (1928-36), Cornish (1902), Fitter (1940), Hampstead Scientific Society (1913), Harting (1889), Johnson (1930), Ministry of Agriculture and Fisheries (1938), Ramsbottom (1943), Ritchie (1931), Robbins (1916, 1927), Willatts (1937).

Plate XXVII

Haymaking in Green Park

Plate XXVIII

Anglers at Highgate No.3 pond with Ken Wood in the background   JOHN MARKHAM

THE INFLUENCE OF SPORT

WITHIN the past two generations the character of the influence of sport on the animals and plants of the London area has radically changed. The emphasis has shifted away from the hunting and killing of certain mammals and birds, known as " game," to the preservation of large areas of open land in the form of playing fields and golf courses, the latter making excellent sanctuaries for wild life.

We saw in Chapter V how, during the nineteenth century, hunting in the neighbourhood of London tended to become a mixture between an open steeplechase and a race meeting. The advent of the motor car still further revolutionised the possibilities of unorthodoxy in the following of hounds, while at the same time the spread of the built-up area reduced the huntable country within fifteen miles of Charing Cross almost to vanishing point. Nevertheless the London foxes throve by raiding the numerous chicken farms that sprang up on the fringe of the suburbs, and recourse had to be made more and more to the sensible solution of shooting them. Not so very long ago foxes destroyed some valuable experimental fowls at the medical research institute at Mill Hill.

There are still quite a number of packs of hounds operating within twenty miles of St. Paul's ; the following list gives the most important pre-war ones, with their headquarters :

Foxhounds

Old Berkeley (Watford, Rickmansworth).
Chiddingfold (Guildford).
Enfield Chace (Hertford, Hatfield, St. Albans).
Essex (Harlow, Ongar, Epping).
Essex Union (Brentwood).
Garth (Windsor).
Hertfordshire North (St. Albans).

Hertfordshire South (St. Albans).
Kent West (Sevenoaks).
Old Surrey and Burstow (Redhill).
Surrey Union (Dorking).

Beagles
Old Berkeley.
Bolebroke (Kemsing, Kent).
Eton College.
South Hertfordshire.
West Surrey and Horsell.
Wolverton (Chigwell, Essex).
Worcester Park and Buckland (Surrey).

Harriers
Aldenham (Herts.).

Otterhounds
Buckinghamshire.
Crowhurst, Sussex.
Eastern Counties.

What the effect of hunting is on the foxes (*Vulpes vulpes*), otters (*Lutra lutra*) and hares (*Lepus europæus*) of the London area, it is difficult to say. Probably it does no more to keep them down than in any other part of the country. Certainly there is less restraint of public opinion on shooting foxes in the London suburban areas than in the shires. The records of the London Natural History Society are full of occasions when punitive shooting parties have been organised to avenge the raiding of local hen-roosts. It was stated in *The Times* of June 22, 1939, for instance, that a fox with a dead rabbit in its jaws had been shot leaving a rabbit enclosure in a garden in Wildwood Rise, Hampstead, and that three foxes had been shot on the Heath within the previous week after rabbits and waterfowl had been missed from Golders Hill Park. In the early spring of 1936 an unsuccessful fox-shoot was held in the Northwood-Ruislip-Harefield district of Middlesex because a fox and two vixens had been raiding local poultry. In Epping Forest it was reported in 1938 that while Hawk Wood was no longer tenable by foxes owing to building development, Bury Wood and several other woods and copses in the Forest still held pairs that produced litters every year. The regular route of the foxes to and from the Forest on their forays lay across the seventh green of the West Essex Golf Course, and the tenant of a cottage near Bury Wood lost six fowls in one week in November, 1938. In the Warlingham and Chelsham district of Surrey it was reported in 1932 that the erection of barbed-wire fences had put a stop to hunting, so that the keepers had to organise fox-shoots, as many as thirty foxes being shot in one season.

From time to time foxes are reported in the London evening papers as having been seen or caught in various parts of Inner London, such as Hyde Park, Kensington Gardens, Brixton and Peckham, but these are almost certainly escapes from captivity. The nearest points to the centre of London where there are still genuine wild foxes appear to be Hampstead Heath, Ken Wood, Mill Hill and Muswell Hill in Middlesex, Epping Forest and Walthamstow in Essex, Elmstead Woods in Kent, and Purley, Wimbledon and Richmond Park in Surrey. At Muswell Hill the proprietor of a chicken run shot a fox in August, 1927, and at Mill Hill a vixen littered in a drain in 1937. At Walthamstow one was seen in a factory yard in Blackhorse Lane on December 31, 1937.

Otters are much scarcer near London than foxes, but are nevertheless commoner than most people realise. They frequent the rivers Lea, Stort, Mole, and probably also the Colne. On the Lea there is a flourishing colony at Walthamstow Reservoirs. (Plate 10b). Members of the Water Board's staff often see one or two, generally at dusk, running along the banks or slipping into the water, and sometimes also find a large fish on the bank with portions typically bitten from the shoulders. In the early 1930's two members of the London Natural History Society, Messrs. R. W. Pethen and E. Mann, saw an otter in broad daylight swimming and diving in the feeder that carries water from the reservoirs to the filter-beds in Lea Bridge Road. This colony, far removed from the possibility of persecution by hounds, has been in existence for at least twenty-five years. In the neighbouring valley of the Roding otterhounds used to hunt the river above Woodford Bridge regularly as late as 1914.

Walthamstow Reservoirs are secluded, but the Thames is far from being a sanctuary, so it is all the more surprising to find that otters have survived there also. Mr. A. Heinemann, writing to *The Field* in 1922, claimed to have seen otters once playing under Westminster Bridge. Mr. C. Grassmann, in the same journal, said he had seen one emerge from beneath a raft off the West End Rowing Club's premises at Hammersmith Mall and enter the river, on November 17, 1923. The correspondence was continued by Mr. H. W. Poulter, who said that there were frequent signs of otters in this neighbourhood, and that one had been seen in a small tidal pool not three miles from Hammersmith in 1921-22. It was presumably from this colony that came the otter which was electrocuted on the railway at Chiswick in 1923, as mentioned in Chapter 11. As recently as August, 1937, an otter was seen on the river bank at Syon Marsh, opposite the Isleworth Gate of Kew Gardens. Strangely enough no records have come to hand of otters seen on the Thames between Richmond and Staines.

The only other creature that is still hunted with hounds in the London area is the hare, which is still common enough in the more rural parts, especially in Surrey. The nearest place to Central London where hares have been seen in the past twenty years are Eastcote, Mill Hill and Scratch Wood in Middlesex, Epping Forest in Essex, Abbey Wood in Kent, and Addington, Wimbledon Common and Richmond Park in Surrey. In Richmond Park there were seven to ten pairs in 1937 ; they usually kept to the plantations by day, but could be found in the open soon after dawn. In the Mill Hill district in 1933-38 hares were seen remarkably close to the built-up area at times. In Epping Forest, on the other hand, they were said in 1939 to have disappeared entirely from the Chingford sector, though ten years previously they were common in the fields next to the West Essex Golf Course at Sewardstonebury. As late as 1911 Webster was able to record that single specimens had been seen in Regent's Park " on rare occasions during late years."

In the nineteenth century hares increased so greatly in Britain that the Ground Game Act had to be passed in 1880 to give the occupier equal right with the landowner to take or kill hares. In some parts of the country this had such a thorough effect that hares had to be protected again by an Act of 1892. In Essex, though Laver in 1898 described the hare as a great pest occurring in all parts of the county, the same author, writing in *The Victoria County History of Essex* in 1903, reported it as " formerly very frequent, but now approaching extinction in some districts as a result of the Ground Game Act." *The Victoria County History of Kent*, however, had quite a different tale to tell, it being reported in 1908 that hares were " nearly as abundant now as formerly, not having been exterminated by the Ground Game

PLATE 33

W. SUSCHITZKY

Azaleas in Kew Gardens

PLATE 34

ERIC HOSKING

Forsythia on Shepherd's Hill, Highgate, looking across to Alexandra Palace

Act, as had been expected." Hares are certainly quite common to-day in the area north and east of Epping Forest, and on the North Downs in Surrey.

Deer will have been remarked as notable absentees from the list of beasts still hunted in the London area. There are no longer any packs of staghounds or buckhounds near London, and the long royal and mayoral tradition has lapsed. Deer, however, are still found in a feral condition in Epping Forest, and are also kept in several parks round London, notably the royal parks at Richmond, Hampton and Greenwich, and private parks at Lullingstone in Kent and Langley in Bucks.

The fallow deer (*Dama dama*) in Epping Forest are the only truly wild deer remaining near London. An annual census is taken by the Forest Conservators, and the herd has remained stationary at about 120 for a good many years past. The deer often stray on to adjoining property, especially gardens and orchards. The indigenous red deer (*Cervus elaphus*) became extinct in Epping Forest in 1827, when the remaining stock was removed to Windsor. A stag and two hinds from the original stock were enlarged in the Forest in the 80's, but proved so destructive that they had to be themselves destroyed. About 1910 a carted red hind escaped from hounds in the Roding valley and took refuge in the Forest, afterwards moving to the Birch Hall estate nearby, from which she made occasional visits to the Forest. Roe-deer (*Capreolus capreolus*) died out in Epping Forest in the sixteenth century, but six were brought from Dorset in February, 1883, and set free in the Forest by Mr. E. N. Buxton, the Verderer. By 1908 there were estimated to be as many as thirty-two in the Forest or neighbouring coverts, but they did not survive long, as the last definite record for the Forest was of one seen at Jack's Hill on August 11, 1917, and they have now died out again. The last record for the Epping district is a doe seen at Lambourne on December 22, 1920, by Mr. Geoffrey Dent. There is good reason to hope, however, that the roe-deer is not entirely lost to the London area, as a doe was seen on one of the more inaccessible slopes of Box Hill early in 1939, at about the same time as a buck was seen at Holmwood, Surrey, a few miles outside the area. All other deer in the London area are kept within park railings.

One other former beast of the chase that is no longer hunted near London is the badger (*Meles meles*), which is still to be found within the County of London, and is much more frequent in the Home

Counties than is generally realised. Being nocturnal beasts, they are rarely seen by diurnal naturalists. The nearest point to Central London and the only place in Middlesex where badgers have been seen within recent years is Ken Wood, where they inhabit a sett in the middle of the wood ; the head keeper saw one twice as recently as 1938, and in September, 1944, a pile of paw-marked sand stood at the entrance of a freshly excavated earth near the duelling-ground. In the early days of aviation a pilot who was guarding his aircraft after a forced landing on the Heath was most surprised to encounter Brock in the early morning, only four and a half miles from Charing Cross. In Epping Forest there are several badger holts, including one at Loughton Camp (Plate 8b) one near High Beech church and one at Theydon Bois. According to Mr. E. N. Buxton, they had not been seen in the Forest for many years when he introduced several pairs in 1886. The other stronghold of the badger near London is Richmond Park, where in 1937 Mr. C. L. Collenette knew of at least four pairs, two inhabited earths being situated at opposite ends of Sidmouth Plantation, and another in Sawpits Plantation. They are accused of having severely depleted the population of mallard ducklings in the park at one time, and for this crime used to be dug out and killed regularly till about 1933, when saner counsels prevailed. On the North Downs in Kent and Surrey the badger still ranks as a common animal, even quite near London. Earths have been reported in recent years from Selsdon, Titsey, Warlingham, Chelsham and Banstead in Surrey, and from Farnborough, Eynsford, and Elmstead Woods in Kent ; a good many others must have gone unrecorded. In South-west Essex, on the testimony of Mr. Geoffrey Dent, badgers are found wherever the soil is right. They are especially common on the chalk around Grays and Tilbury, and also occur at Ongar and South Weald, and in Hainault Forest.

The influence of the sport of shooting on the bird population is the same in the London area as in other parts of Lowland England, except that there is probably less full-scale game preservation in the outskirts of London than elsewhere. The three game-birds concerned are the pheasant, the common or English partridge, and the red-legged or French partridge. Pheasants are the most artificial of the three, as not only are they introduced, though of long standing, but they would probably not survive if their predators were not ruthlessly repressed. The common partridge, the only indigenous bird of the three, may still be found within a dozen miles of St. Paul's in such places as Richmond Park and Croydon Aerodrome, where it is protected for non-sporting reasons. To-day, in the words of a well-known pre-war advertisement, " there are no partridges in Piccadilly," but in April, 1939, stray birds appeared in small gardens in Bayswater and Fulham. The red-legged partridge, introduced to England by eighteenth century landowners, has always had its stronghold in Eastern England, and Essex is at present the Home County where it is commonest. " Frenchmen " are reported annually as near London as Chigwell and Loughton in the Roding valley, and at Sewardstone in the valley of the Lea twenty-one were shot in one day in October, 1942. One other game-bird, the quail (*Coturnix c. coturnix*), is very exceptionally recorded near London, the most recent occasions being at Great Parndon, Essex, in May, 1942, and at Cheam, Surrey, in June, 1938. Live quails were at one time imported for the table, and probably some escaped, for they occasionally turned up in suburban back gardens, for instance at Kilburn about 1919, on the evidence of Mr. Johnson.

The most important influence of game preservation has not been on the preserved game-birds so much as on their predators, the birds and beasts of prey, including many that normally never look at a game-chick. In the London area, as in the rest of Southern and Eastern England, such fine carnivores as the polecat and pine-marten have long been extinct, and many birds of prey, notably the kite and buzzard, have also been driven out.

The last polecat (*Mustela putorius*) in South-west Essex was killed in Epping Forest in the early 90's, the last in South Herts in Oxhey Woods in 1872, and the last in North-east Surrey at Headley Park about 1886. One that was shot on Wimbledon Common in 1936 had almost certainly escaped from captivity. The last pine-martens (*Martes martes*) near London were one shot in Redlands Wood, near Holmwood, Surrey, in 1879, and one seen in Epping Forest near Ambresbury Banks in 1883. It must be said in defence of the pheasant, if it is accused of having cost us the polecat and the marten, that the large number of chicken-farms round London makes it very doubtful whether the continued existence of these two handsome but blood-thirsty carnivores could in any event have been tolerated.

The neighbourhood of London once boasted eight or more species of diurnal raptorial birds, the hobby, kestrel, common buzzard, three harriers (*Circus æruginosus, C. pygargus, C. cyaneus*), sparrow-hawk and kite (*Milvus milvus*). Of these, thanks very largely to the keeper's gun, only the kestrel and sparrow-hawk remain, both of them happily still tolerably common. The chief blame for the disappearance of the harriers must, however, be apportioned to the draining of the marshes where they breed. The hobby has not been recorded breeding in Middlesex since 1861, when three eggs were taken from an old crow's nest in Pinner Wood. A pair nested at Belhus Park, Essex, within the twenty-mile radius, in 1879, and another was seen near the same locality in 1910. In August, 1864, a female hobby was killed on Hampstead Heath while pursuing a wounded swallow.

Three or four hundred years ago the kite was the bird of prey that had adapted itself to urban life in London. To-day its place has been taken by the kestrel, which is probably commoner in the suburban fringe where it is not persecuted than in the rural fringe where keepers still wantonly shoot it in defence of the privileged pheasant, in spite of the fact that it is protected, and that 64.5 per cent of its food consists of mice and voles and a further 16.5 per cent of injurious insects. In fact only 8.5 per cent of a kestrel's diet, according to Dr. Collinge, consists of birds, and this is mainly composed of small song-birds. It is only the exceptional individual kestrel which preys on pheasant or partridge chicks, and there appears to be no record of it attacking any adult game-birds. The abundance of rats and mice in Inner London means that several pairs of kestrels can live in the heart of the metropolis ; sometimes they take house-sparrows as well. Within recent years kestrels have almost certainly nested annually on the towers of Westminster Abbey (or the adjacent Victoria Tower of the House of Lords) and the Imperial Institute in South Kensington. Kestrels would be a much commoner sight in the London sky if gazing upwards were not such a dangerous occupation in London's traffic-laden streets. A kestrel hovering over the Horse Guards Parade is by no means so unusual as might have been assumed from a recent correspondence in *The Times*, which ended with a letter from an observer who said he feared " the " kestrel was now dead, as he had seen it fluttering down dying from his hotel into the street.

In the case of the sparrow-hawk the animus of the game-preserver is more justified, as its diet includes 16.5 per cent of game-birds and 16 per cent of poultry and ducks, according to Dr. Collinge. In contrast to the kestrel, two-thirds of its diet consists of birds. The sparrow-hawk is much less of a London bird than its rival, though at least one

Plate XXIX

A giraffe at the Zoological Gardens in Regent's Park   JOHN MARKHAM

Plate XXX

Little owl with cockchafer at nest-hole   ERIC HOSKING

pair still frequents the Hampstead and Highgate district, and it is not uncommon on the rural fringe. In 1943 sparrow-hawks were four times recorded in Inner London, and twice were seen to swoop on house-sparrows.

The smaller carnivores and the owls also suffer from the attentions of the gamekeeper. Stoats and weasels are still, however, quite common in the more rural parts of the London area. Though they relish a rat as much as any other meal, they have not chosen to follow the rats into their urban fastnesses in the sewers and warehouses. Stoats (*Mustela erminea*) come as near to London as Ken Wood and Mill Hill in Middlesex, Epping Forest in Essex, Abbey Wood in Kent, and Wimbledon and Richmond Park in Surrey. Weasels (*Mustela nivalis*) have a very similar distribution.

The killing of owls in the interests of game-preservation is quite as unjustified as the killing of kestrels. Unfortunately the issue has been clouded by the controversy over the little owl (Plate XXX), an intruder from the Continent released in Kent and Northamptonshire by Lord Lilford and others at the end of the last century, which has since colonised practically the whole of England and Wales south of the Humber. The first breeding record for the little owl in Middlesex was in 1912, when a nest was taken at Enfield, but the bird has never penetrated far into the London suburbs as a breeder. Though the inquiry sponsored by the British Trust for Ornithology in 1936 has definitely proved that the little owl is a beneficial bird, since its diet consists predominantly of insects and small mammals, a die-hard party among the game-preservers persists in citing individual cases of little owls which have taken to a diet of game-chicks in order to damn the whole race.

Unhappily the little owl controversy has helped to reinforce the prejudice against all owls that still exists among old-fashioned game-preservers. The comments of Mr. G. B. Blaker, who conducted the Barn Owl Inquiry of 1932 for the Royal Society for the Protection of Birds, are interesting in this connection :

" Information was forthcoming even from gamekeepers themselves, some of whom admitted having shot Barn Owls, while others shoot them without admitting it. There still exists that section of the gamekeeping community which has been brought up to respect the old tradition of their ancestors which says that anything with claws and a hooked beak is a deadly enemy. But this section is now a minority, and there is a healthy tendency growing among game-keepers to regard the Barn Owl as a valuable check on the smaller and more elusive types of vermin. . . . Most landowners, also, were very helpful, and in one or two cases the interest aroused by the inquiry has resulted in instructions to keepers that these birds are to be left alone in the future. Unfortunately, the irre-sponsible man with a gun still remains a difficult problem ; and the tendency, brought about by modern conditions, for owners of big shoots to let them to syndicates which have no real interest in the land over which they shoot is most unhealthy from the Barn Owls' point of view."

The barn-owl now unfortunately ranks as a scarce bird near London, due perhaps less to the activities of gamekeepers than to other human factors noted by Mr. Blaker, including the wiring up of many church towers and belfries to keep out the jackdaws, an act which also deprives the owls of a favourite nesting haunt. Few modern buildings have the openings in the roof where the owls like to build their nests, and which were a feature of the old barns which gave the owls their name. Many old trees with suitable nesting holes have fallen before the advance of London.

The true owl of London is the tawny or brown owl, which is resident in most of the central parks, and can often be heard in the quieter streets at night. It was of the tawny owl that Alfred Noyes wrote :

" The linnet and the throstle too, and after dark the long halloo,
    And golden-eyed tu-whit, tu-whoo, of owls that ogle London."

Fire-watchers have seen these London owls hunting the streets for rats and mice, and dwellers in Bloomsbury may have difficulty in distin-guishing between the cries of hunting owls and the hooting of the Scottish expresses leaving Euston Station. At the open air theatre in Regent's Park, too, the tawny owls, together with the distant yahooing of the Zoo gibbons, provide a most apt accompaniment to parts of *The Midsummer Night's Dream* or *The Tempest* :

" There I couch when owls do cry.
    On the bat's back do I fly . . ."

(Bats are quite likely to be flying overhead during this passage, as well as owls hooting). A most vociferous calling of tawny owls is punctuating the writing of this chapter, in a flat at the summit of Highgate West Hill.

One of the most important effects of game-preservation is an indirect one, the large increase in the number of small mammals and birds made possible by the absence of their natural predators. Though we lament the lack of hawks and falcons, polecats and martens, many would consider themselves more than adequately compensated by the choruses of blackbirds, thrushes, warblers and other song-birds. Unfortunately there has never been a proper study of the balance between raptorial birds and their prey, so that we do not know how an area where birds of prey are severely persecuted differs from a similar one where there is little or no keepering, except in a very rough and ready way. The suburban area of London cannot be taken as typical in any respect. Though there is less persecution of birds of prey near London than elsewhere, only the kestrel has really adapted itself in any degree to urban conditions.

Besides the shooting of partridges and pheasants, there has always been a good deal of more or less rough shooting of rabbits and hares, and such minor game-birds as woodcock, snipe and wild duck, in the London area. This also has declined in importance, for the area of open country where it is possible to take out a gun and have pot-shots without the danger of peppering a courting couple behind a hedge is continually decreasing near London. Cornish relates how a man he knew used to shoot snipe on Chiswick Eyot, and wild duck and teal on a stream running from Chiswick House to the Thames, while another had broken a young pointer to partridges on a market garden between Barnes Bridge and Chiswick. This sort of sport is now almost impossible within about fifteen miles radius of Charing Cross.

In mid-Victorian times it was considered "sport" to shoot at any unfortunate bird that came within range, as is well illustrated by Charles Kingsley's description of

"a purposeless fine-weather sail in a yacht, accompanied by many ineffectual attempts to catch a mackerel, and the consumption of many cigars ; while your boys deafen your ears, and endanger your personal safety, by blazing away at innocent gulls and willocks, who go off to die slowly."

A similar callousness is evident in an incident related by W. H. Hudson :

> " That was a memorable season (1892-93) in the history of the London gulls. Then, for the last time, gulls were shot on the river between the bridges, and this pastime put a stop to by the police magistrates, who fined the sportsmen for the offence of discharging firearms to the public danger."

That these were not isolated instances of thoughtless cruelty is shown by the attitude of mind of a cabby whom Pigott once saw strike sharply with his whip at a yellowhammer feeding in the gutter—one would have thought that the unusual presence of a yellowhammer in London deserved some more humane recognition. To realise how greatly public opinion has changed in this respect, it is only necessary to think what would be the reaction of passers-by to a cabby seen committing such a barbaric act to-day. Two incidents related by Mr. Johnson indicate that this change has only become completely effective since the end of the 1914 war. When the rooks attempted to re-colonise the Inner Temple in the spring of 1916, they were frightened away by somebody shooting at them with an air-gun. In May, 1918, a bottle-nosed dolphin (*Tursiops truncatus*) came up the Thames as far as Battersea, and was wounded by people shooting and throwing missiles at it, so that it was stranded and died.[1]

The sport of angling still engages many hundreds of Londoners, especially along the banks of the Lea (Plate 25), and in such popular resorts as the Serpentine, the Hampstead and Highgate Ponds (Plate XXVIII) and the lake in Battersea Park. Here carp, bream, roach and perch may be caught, and in 1926 a jack weighing six pounds was taken in the Vale of Health Pond on Hampstead Heath. In heavily fished waters angling is so important a biotic factor that they have to be periodically restocked, and the whole position is highly artificial. The disappearance of trout and perch from the Wimbledon Common meres is, however, attributed by Mr. Johnson to the depredations of herons from Richmond Park.

The comparatively recent tendency for large areas of open land to be preserved in the form of playing fields (Plate XXIII) and golf courses (Plate XXII) has had an important effect by creating extensive islands of grassland in the middle of otherwise built-up areas. It is

[1] Cf. also Richard Jefferies's essay, " The Modern Thames " in *The Open Air* (1885).

difficult for people reared in this age of golf and football to realise how very recent this development is. Though the first golf course in the London area was made on Blackheath in 1608, there were no others till the London Scottish on Tooting Bec Common in 1863, and the Royal Wimbledon on Wimbledon Common two years later. The real fashion for golf began in the late '80's, and by 1900 there were already fifty-one golf clubs in Middlesex and fifty-five in Surrey. Several of the older courses have succumbed to the builder, and by 1939 there were only thirty-eight in Middlesex, out of some 140 in the whole London area.

An interesting by-product of the laying out of golf courses has been the transplanting of maritime plants along with the seaside turf used for the construction of greens. The sea milkwort (*Glaux maritima*) and the sea-plantain (*Plantago maritima*) have been found growing freely on and near golf greens on Wimbledon Common, having evidently been introduced with turves. The marram grass (*Ammophila arenaria*), a lover of maritime sandhills, was planted on bunkers on Northwood golf course in 1913.

The trampling, rolling and mowing of playing fields destroy all the taller plants, leaving to a large extent pure grassland. On golf courses, on the other hand, there are many patches of rough where the original herbage continues to grow undisturbed. Of the birds that feed on golf courses and playing fields, the most important are the gulls and starlings, though many thrushes, lapwings and other birds also frequent them. Mr. E. C. Rowberry has told how the gulls feed only in the early morning before the sun is properly up :

" In order to see gulls taking toll of vast numbers of earthworms on these fields, it is necessary to be up at dawn, when the first birds commence to straggle in from the western reservoirs. On September 18th, 1933, I made the following observations on a golf-links at Osterley. About 5.30 a.m. Lesser Black-backs commenced to arrive ; they alighted and commenced to feed at once. Carefully watching one bird, I noticed that it devoured forty-five worms in five and a half minutes. Another bird took twenty-six in three and a half minutes. At 6.10 a.m. there were 254 birds of this species feeding. If we assume that this number of birds feed for only fifteen minutes, at the rate of ten worms per minute, we reach a total of 38,100."

References for Chapter 16

Baker (1908), Beadell (1932), Bishop *et al.* (1928-36), Buxton (1901), Bucknill & Murray (1902), Blaker (1934), Collenette (1937), Collinge (1927), Committee on Bird Sanctuaries in Royal Parks (England) (1929-39), Cornish (1902), Crossman (1902), Dawson (1940), Forbes (1911), Glegg (1929), Harting (1866), Hibbert-Ware (1938), Hudson (1898), Johnson (1930), Kingsley (1855), Laver (1898, 1903), London Natural History Society (1900-44, 1937-44), Marriott (1925), Pigott (1902), Ritchie (1931), Rowberry (1934), Sinclair (1937), Stubbs (1917*a*), Webster (1911), Witherby *et al.* (1938-41).

ADDENDUM : Fairway and Hazard (1939).

THE CULT OF NATURE

" Go down to Kew in lilac-time, in lilac-time, in lilac-time ;
Go down to Kew in lilac-time (it isn't far from London !)"
ALFRED NOYES.

THE relationship between man and the other natural communities, both animal and plant, with which he shares this planet have always been somewhat ambivalent. In the foregoing chapters many aspects have been described of man's competition, both conscious and unconscious, with the other communities in the lower Thames valley. So long as the Thames was fringed with alder swamps, there could be no London, so the swamps had to be drained and the alders to go. So long as the site of London was girt about with a dense oak forest, there could not be enough agriculture to support a large human community, so the forest had to be destroyed, and with it the larger beasts which it harboured. Again and again, we have seen the disappearance of some animal or plant community, either because it competed directly with man's needs, or because the changes required for man's well-being were incompatible with its survival.

Yet, all the while, there has been a strong feeling of kinship, friendship, love, admiration, liking, call it what you will, for the lower orders of creation on the part of the dominant mammal, man. It appears in the primitive pantheistic religions, which saw spirits in beasts and birds and trees ; it appears in the totemic religions, with their transference of group personality to some class of animal or plant ; it appears even in our own times in such curious mixtures of feeling as that of hunting Englishmen towards the foxes they hunt, which has led to such widespread accusations of hypocrisy from abroad.

Thus the modern Londoner has a most complex web of attitudes towards the animals and plants that have succeeded in adapting themselves to life in or near his great urban aggregation. Some, such as cats and dogs, actually share his home, and are the objects of real affection and feelings of companionship ; others, such as chickens and rabbits, are kept and killed for food, satisfying both material needs and the deep-seated primitive human urge to kill that which is familiar and loved ; still others, such as rats and mice, flies and lice and bugs, are regarded as unmitigated pests, and ruthlessly extirpated whenever possible. Out of doors, there is the same tendency for some wild creatures, such as gulls in the parks and tits and the robins in the gardens, to be treated as pets, while others, like foxes and deer, are regarded ambivalently, and yet others, such as cabbage butterflies, wireworms and greenfly are destroyed without any *arrière-pensées*. Towards plants, similar trends are discernible. There are, of course, in many if not most cases material as well as sentimental reasons for these attitudes ; for instance flies and lice are carriers of disease as well as sensually repulsive, and robins are beneficial to the gardener as well as æsthetically pleasing.

Besides these various attitudes towards individual animal and plant species, many people who live in completely urban conditions feel a great urge to renew contact with the country from time to time. Much nonsense has been written about every Englishman being a country-man at heart and longing only to retire to a cottage in the depths of the country where he can lean over the wall of the pigsty and scratch the pigs on the back. The truth is, as Professor D. W. Brogan has recently pointed out in his able exposition of the English character for the benefit of American readers, that

> " there is no country in the world in which feeling for the soil as a *factor of production* is as rare as in England, or where knowledge of farming as a way of making a living is so much a specialist know-ledge, or in which the most romantic and unrealistic views of country life can be advanced with less danger of brutal contradic-tion from people who know what agriculture, as an economic and social system, involves."

What is true of England as a whole is *a fortiori* even truer of London. The illusion that all cockneys are pining for the delights of rural life has largely been exploded by the experience of the evacuation in 1939, when no group came back to their homes more quickly than the London mothers, bored to death by a combination of nothing to do and the quiet of the country. The main cause of the illusion is that, to quote Professor Brogan again,

> " two different things are confused, the love of flowers, gardens,

Plate XXXI

Cormorants in St. James's Park

Two mallard on planks floating in a static water tank in London

Plate XXXII

Oxford ragwort on a blitzed site near St. Paul's   JOHN MARKHAM

open spaces, and the holiday delights of rural life, with the less picturesque, less common, less literary passion for the utilisation of the land that marks the true peasant."

What is undeniable, and has important consequences for the natural history of London, is that very large numbers of Londoners, particularly the younger ones, find continuous life in a wholly built-up area so unendurable that they must refresh themselves in the country at week-ends. Hence the crowded trains and buses on Sundays to such popular resorts as Box Hill, Epping Forest, and the Chilterns.

Among animals, the principal indoor pets, which have the run of the house in most cases, are dogs and cats. The dog population of Great Britain increased by more than 50 per cent in the ten years before the war, and in 1938-39 over three million licences were taken out. Due to the war, the number fell to a little over 2,500,000 in 1941-42. The number of dog licences in the County of London in 1937-38 was 204,288, or roughly one for every twenty human inhabitants. The most important biotic influence of dog-keeping is probably on agriculture rather than on the wild animal communities. In many districts round London sheep-rearing has had to be abandoned because of the large number of uncontrolled dogs that worry the breeding ewes.

Cats, on the other hand, form quite a significant control on the bird population of the central and suburban districts (Plate XXIVb).[1] W. H. Hudson was much exercised about "the cat question." He estimated in 1898 that there were nearly three-quarters of a million cats in London, compared with not more than two or three hundred thousand dogs. Hudson thought that only this large army of cats kept down the numbers of the London sparrows, which he estimated at two to three millions. Cats are especially dangerous to sparrows when they are just leaving the nest, but they are probably an even more important factor in keeping down small birds that nest in less inaccessible places than sparrows, such as blackbirds, thrushes, robins, wrens and finches. Though the scarcity of these in Central London is undoubtedly largely due to the lack of insect food, in turn largely due to the soot encrusting the vegetation, the excessive number of cats, always ready to eke out

---

[1] Cf. an interesting recent paper by Colin Matheson, Head of the Department of Zoology, National Museum of Wales, on " The Domestic Cat as a Factor in Urban Ecology " (*J. Anim. Ecol.*, *13*, 130-33, November, 1944). The percentage of school children's homes in Newport (Mon.) and Cardiff containing cats ranged from 26 per cent to 76 per cent in different schools.

their dustbin diet with young birds, is an almost equally important control. It is significant that the black redstart, which has begun to colonise Inner London in the past five years, is a bird that usually nests in sites far more difficult of access to cats than the trees and shrubs favoured by most of the birds in the suburban garden association. The ornamental water birds in London also suffer from cats. In Hudson's day one cat killed and partly devoured two sheld-ducks and a tufted duck on one of the islands in the lake in Battersea Park, and others succeeded in exterminating a colony of rabbits that had been introduced into Hyde Park.

Other animals kept indoors as pets include birds in cages and fish in aquaria. Before the enforcement of the Wild Birds Protection Acts, large numbers of goldfinches and other small birds were taken every year in the country round London to meet the demand for cage-birds. In the autumn of 1865, for instance, Harting saw several siskins which had been taken near Hendon in company with lesser redpolls by a bird-catcher. The occasional cases of siskins nesting in southern England are usually attributed to escaped cage-birds. Nightingales were also much persecuted by bird-catchers round London, though it is not clear whether they were taken for the pot or for caging. Harting knew a Middlesex gamekeeper who paid his rent by the capture and sale of nightingales, of which in one season alone he disposed of 180 for eighteen shillings a dozen in London. After that, it is not surprising to read Harting's verdict that "it is now by no means so common a bird in the county as formerly."

Cage-birds not infrequently escape, but the exotic ones rarely survive long, and none have established themselves as resident British species. Canaries (*Serinus canarius*) and budgerigars (*Melopsittacus undulatus*) are probably the two cage-birds most commonly kept in the London area, and so are the two most often seen at large. The author once saw a budgerigar in Golden Square, off Regent Street, that was being mercilessly harried by sparrows, and has five times seen them at large in suburban Surrey ; in August, 1932, one was accompanying a flock of about 150 sparrows on the stubble at Riddlesdown near Purley, and was apparently not being molested. In 1938 two aviary escapes were seen in St. James's Park. One was an exotic species of ring-dove that frequented Duck Island from June to the end of the year, and the other a rusty blackbird (*Euphagus carolinus*), a North American species, which stayed a month, often accepting food from human hands, and was eventually found drowned. Ritchie mentions that about 1905-07 a number of exotic doves, such as bronze-wings, turtles and crested pigeons, together with Pekin robins (*Liothrix luteus*) were released in London as part of a definite attempt at acclimatisation, but without success.[1]

Fishes and other inmates of aquaria cannot escape in the accepted sense of the term, but they nevertheless get released from time to time. Horn recorded two goldfish (*Carassius auratus*) from the docks during the drought of 1921, which could only have come from captivity. Small water tortoises, probably the European water tortoise (*Emys orbicularis*), have occasionally been seen in ponds near London, and as they are not indigenous in this country can only have escaped or been released. One was seen at a gravel-pit at Beddington, Surrey, in April, 1933, and another was basking on a log in the Upper Pen Pond in Richmond Park in July, 1934. It is on record that six full-grown tortoises of this species were enlarged in a small artificial pond at Shere, Surrey, about 1890 ; two wandered and were brought back ; two survived till 1906, when two more were put into the pond.

Several attempts have been made to establish the edible frog (*Rana esculenta*) in the London area—a somewhat questionable form of activity in view of the vocal accomplishments of this frog, which have earned it in other parts of the country, where it has been established, the names of " Cambridgeshire nightingale " and " Whaddon organ." One of the earliest attempts was that of the naturalist Doubleday at Epping, though none seem to have survived to the present century. About fifty years ago, when experiments in acclimatising strange and often undesirable creatures in Britain were rather fashionable, numbers of this frog were imported from Belgium and Germany and released at Shere and Chilworth, whence they escaped to Gomshall Marsh ; fifty more were imported from Berlin to the same locality by Lord Arthur Russell in 1894-95, and three or four were still surviving in 1906. In August, 1904, a small pond at Ockham, about six miles north of Shere, was said to be alive with them. By the 1930's edible frogs were well established in several localities in Surrey and Kent within twenty miles of London. In June, 1931, Mr. J. Rudge Harding, the official bird-watcher for Richmond Park, saw and heard one " singing " in Bishop's Pond in the Park ; the song resembled the

[1] I am indebted to Prof. Ritchie for providing me with the reference to this experiment (Finn (1905)), which shows that some four dozen Pekin robins were deliberately released in St. James's and Regent's Parks in April, 1905.

" hammering of a machine-gun." A less successful centre of introduction was near Kentish Town, in the garden of Meadow Cottage
in Highgate Road, which had already been converted into a playing
field by 1913.

Though it is not strictly natural history, a reference to the largest
collection of wild animals in London, the Zoo in Regent's Park, is
perhaps not out of place here (Plate XXIX). This most popular of
London's outdoor entertainments is provided for the masses on land
leased from the Crown in a Royal Park by one of the most
patrician of British scientific societies. The Zoological Society
of London is one of the few major learned societies lacking
the epithet " royal." In the thirty-four acres of the Society's famous
gardens, visited by 2,339,357 people in 1938, the last full year of
peace, there were 3624 animals of all kinds (excluding fishes and
invertebrates) on December 31 of that year. These animals ate an
impressive quantity of food each year in peace-time, ranging from
91 tons of hay and $156\frac{1}{2}$ tons of clover, to 124 tons of horse-flesh, nearly
a quarter of a million bananas and $4\frac{3}{4}$ cwt. of dried flies. In addition
to the many popular favourites, such as lions (successors to the historic
lions in the Tower that formed the nucleus of the collection), tigers,
polar bears, elephants, giraffes, parrots and monkeys, the society has
imported to London and successfully kept many of the rarest and most
delicate animals in the world. Thus within the pre-war years it was
possible to see within two miles of Charing Cross such rarities as the
gorilla and okapi of the Central African jungles, the giant panda of
the remote highland forests of innermost Asia, the first snub-nosed
monkey, another denizen of Western China, ever to be exhibited in
any zoo, the first specimens of the gorgeously plumaged quetzal of
Ecuador to reach Europe alive, and a collection of the fragile humming-
birds of tropical South America.

The London Zoo is less important as a natural habitat than as one
of the central shrines of the animal cult. Generations of Londoners
would hardly have known what an elephant or a lion looked like, if
they had not been able to see them in the flesh, first in the Tower and
later in Regent's Park. Even as a natural habitat, however, the Zoo
is not negligible. It harbours considerable colonies of such alien pests
as rats, crickets and cockroaches. At least one animal, the grey squirrel,
has escaped or been released from it and settled the surrounding
terrain, while many others, especially birds, have escaped without

PLATE 35

W. SUSCHITZKY

Bluebells in Oxhey Woods, Herts.

PLATE 36

ERIC HOSKING

Highgate Wood, showing the fenced-off bird sanctuary

establishing themselves. At the beginning of the war a number of birds of prey, including black kites (*Milvus migrans*), were released, and one of the black kites was later seen at Stratford in the East End. During the blitz a zebra and a cockatoo, among other things, escaped ; the zebra was recaptured after an exciting chase round the Outer Circle of the Park, but the cockatoo remained at large ; one of the ravens, whose cage was destroyed, was recaptured next day. Innumerable house-sparrows inhabit the Zoo and its various houses, hopping about in the presence of the fiercest beasts and birds with the greatest fearlessness ; they are even found inside artificially warmed places, such as the monkey house.

The Zoo being partly indoors and partly out of doors, leads us on to consider the keeping of pets out of doors. The relationship between man and his outdoor pets is, with the single exception of the horse, less intimate than with those beloved commensals the cat and dog. In London few horses are now kept purely for riding, and compared with the nineteenth century multitudes few even for draught purposes. Indeed, the noisomeness of the fumes to which draught horses are subjected in the petrol-laden air of the London streets to-day suggests that it would be a kindness to the horses, as well as a great boon to other traffic, to ban them altogether from the streets of London. From a biotic point of view the chief importance of the horse is that its droppings and its nose-bag chaff provide some of the staple food of the ubiquitous house-sparrow (Plate XVa).

Next to horses, deer are the largest animals kept in captivity outside zoos in the London area. Up to 1914 there were deer in quite a number of parks, both private and public. Red deer were kept in Richmond Park, Surrey, and Langley Park, Bucks. Fallow deer were kept in Ashtead, Battersea, Carshalton, Morden Hall, Richmond and Wimbledon Parks in Surrey, as well as in Greenwich Park and Langley Park. Since 1920 the number of parks near London with deer has decreased considerably, and fewer deer have been kept in the parks that still have them. Richmond Park still has both red and fallow deer, and so does Bushy Park a few miles away on the other side of the Thames. Battersea Park in South London has four red and six fallow deer in a small enclosure ; Golders Hill and Clissold Parks in North London also have a handful in enclosures. Greenwich and Lullingstone Parks in Kent, and Langley in Bucks, still have their deer, but many of the great herds are gone. Keeping deer in a park is an expensive pastime,

and only public bodies and very wealthy landowners still have the
resources for providing this form of free public entertainment.

The herd of deer in Richmond Park consisted of 50 red and 1600
fallow deer in the early part of the nineteenth century. The numbers
of red (Plate 26a) remained steady for a hundred years, but by 1937
had risen to 80, the highest figure recorded since the seventeenth
century. The number of fallow, however, fell gradually to about 1150
in 1892, then to 900 in 1909, and finally to only 300 in 1937. At one
time small numbers of roe deer were kept in the plantations in the
park, and one or two were occasionally seen in Sawpit Wood till about
thirty-five years ago, but they seem to have died out since.

Two of the most successful examples of the rage for introducing
alien animals to the British Isles that overtook landowners at the end
of the nineteenth century are the grey squirrel and the little owl. The
spread of the little owl from its Kentish centre of introduction was
described in Chapter 16.

Between 1890 and 1916 grey squirrels (*Sciurus carolinensis*) (Plate
XVb) were liberated, with exactly what motive it is difficult to discover,
at half a dozen places within twenty miles of St. Paul's, and at one
place only a mile or so outside the London area. By 1930 these natives
of North America had colonised the whole of Greater London, except
for Essex, which was isolated by the rivers Thames and Lea, and an
area in western Middlesex, which is in any case unsuitable country.
There is no evidence whether the first grey squirrels released near
London, five set free in Bushy Park, Middlesex, by Mr. G. S. Page of
New Jersey in 1890, survived or not, but when a hundred were enlarged
at Kingston Hill, Surrey, a few miles away, by another American
citizen a dozen years later, he was, perhaps unfortunately, successful
in his aim. Between 1905 and 1907 ninety-one grey squirrels were
brought from the Duke of Bedford's estate at Woburn, and released in
the Zoo or in Regent's Park, whence they soon spread to adjoining
open spaces. Two secondary centres of introduction were at Farnham
Royal, Bucks., in 1908-09, and Sevenoaks, Kent, in 1911.

From these places the squirrels spread, fairly slowly up to about
1920, but rapidly thereafter. From Kingston Hill they colonised
western Surrey and southern Middlesex, reaching Bushy Park about
1903, Molesey in 1909, and Roehampton before 1911. By January,
1923, there were estimated to be as many as 150-200 grey squirrels in
Richmond Park. After 1920 they advanced rapidly along three main

routes, southward along the Mole valley, westward up the Thames valley and eastward through the parks and commons of South London. In 1923 one was seen swimming the Thames near Eel-pie Island. Eastern Surrey and western Kent were colonised from the Sevenoaks centre ; by 1926 these squirrels had occupied most of the North Downs area.

From Regent's Park, grey squirrels spread to Hampstead in 1908, and to Hyde Park at about the same time ; by 1913 they had thoroughly occupied the surroundings of Hampstead Heath. Soon after 1920 they began to appear in squares and gardens all over Inner London, pene-trating as near the central treeless core as Russell and Woburn Squares. In 1923 there were estimated to be 250 grey squirrels at large in Regent's Park and about a score in Hyde Park and Kensington Gardens. The eastward spread of the Regent's Park squirrels, how-ever, was slow, and they did not cross the Lea to make a permanent colonisation of the Epping Forest area till about 1936. Since then their progress has been rapid, and they are now almost as common in the Forest as anywhere else round London. Northwards the line of advance from Regent's Park met the eastward thrust from the Colne valley. The Farnham Royal squirrels were also slow off the mark ; none were reported across the Colne till 1919, when one appeared in Oxhey Woods, Herts. In the succeeding ten years the whole of northern Middlesex and southern Hertfordshire were colonised by squirrels originating either in Regent's Park or at Farnham Royal.

The present status of the grey squirrel in the London area is that of an all too common pest, for it is almost omnivorous, eating most things from young birds and green shoots to the leaden labels on the trees in Kew Gardens. Some idea of their abundance can be gained from the fact that it was estimated that some 4000 had been shot in Kew Gardens in the twenty years ending 1937. War was declared on them in the Royal Parks in 1930, and in the London County Council Parks in the following year. In Richmond Park 2100 were shot between 1932 and 1937, and in the latter year there were believed to be still five or ten pairs in the park, while on the nearby Wimbledon Common over 200 were shot in the last few months of 1937. In the four Central Parks some 170 grey squirrels were shot within a few weeks of the extermination order, and in 1938 only one was reported from Ken-sington Gardens. Over 300 grey squirrels had been shot in Epping Forest up to 1942. In the open spaces controlled by the L.C.C. some

300-400 grey squirrels were shot annually between 1932 and 1937, with the result that by 1937-38 only four open spaces under the Council's control held any of the pests ; these were Hampstead Heath and surrounds (132 shot), Marble Hill, Twickenham (58 shot), Beckenham Place Park (102 shot), and Castlewood and Jackwood (1 shot). Though there were still grey squirrels to be seen in Ken Wood in 1944, it can be said that the war on them has been fairly successful.

It is often maintained that the grey squirrel has driven out the red squirrel over a great part of England. The truth seems to be that the grey arrived in this country at a time when the fortunes of the red were at a very low ebb, and was thus able to occupy the niche left vacant by the red's decline following a disastrous epidemic about fifty years ago. It must be remembered that the natural habitat of the red squirrel is coniferous woodland, and it was only because of the intensive persecution of its natural enemies, especially martens, by game-preservers, that it was enabled to spread into the deciduous woodlands in the south in such abundance during the nineteenth century. At the peak of its expansion, round about 1900, the red squirrel was smitten by one of those mysterious and catastrophic epidemics that happen in the animal world just when a species seems to be going strong, and lost nearly all the ground it had gained in the previous hundred years or so. In the London area, where it had been quite common in the outer wooded districts, it became almost extinct, except in the neighbourhood of Epping Forest, with the result that the grey was able to step into its shoes without a fight. In the past ten years, however, the red squirrel seems to be spreading once more, especially in Surrey and within a few miles of Epping Forest.

The explanation of the survival of the red squirrel in the Epping Forest area throughout the present century is a very interesting one. The indigenous red squirrel (*Sciurus vulgaris leucourus*), according to Stubbs, was decreasing in Essex after 1909, and was virtually extinct in the western part of the county by 1917. About 1910, however, an Epping landowner, Mr. C. E. Green, bought a number of Continental red squirrels (*Sciurus v. vulgaris*) in Leadenhall Market and released them on his estate. Stubbs considered that the red squirrels to be seen in and about the Forest from 1917 onwards were really of this form, and not of the original English type, and this theory has been borne out by the identification as Continental of a skin picked up in the Forest in 1936. At all events the red squirrel has flourished in Epping Forest

PLATE 37

Rosebay willow-herb and Canadian fleabane in a ruined
City church   ERIC HOSKING

PLATE 38

Coltsfoot on a blitzed site   ERIC HOSKING

and languished over most of the rest of southern England, so that something must have happened to exempt them from the scourge that overtook their congeners.

Many exotic species of waterfowl are kept on artificial lakes both in and around London. In addition to the Crown collections in the Central Parks and Kew Gardens, there are private collections on many large estates, such as Trent Park, Middlesex, Foxwarren, Surrey, and Woburn Abbey, Bedfordshire.

The following complete list of the waterfowl kept at Kew Gardens in 1939, kindly supplied by the Director of the Royal Botanic Gardens, contains most of the swans, ducks and geese kept on artificial waters near London :

Black-necked Swan (*Cygnus melancoriphus*).
Black Swan (*Chenopsis atrata*).
White-fronted Goose (*Anser albifrons*).
Bar-headed Goose (*Eulabeia indica*).
Chinese Goose (*Cygnopsis cygnois*).
Barnacle Goose (*Branta leucopsis*).
Canada Goose (*B. canadensis*).
Andean Goose (*Chloëphaga melanoptera.*)
Magellan or Upland Goose (*C. leucoptera*).
White-faced Tree-Duck (*Dendrocygnus viduata*).
Fulvous Tree-Duck (*D. bicolor*).
Muscovy Duck (*Cairina moschata*).
South African Sheld-Duck (*Casarca cana*).   (Plate 21b).
Paradise Sheld-Duck (*C. variegata*).
Common Sheld-Duck (*Tadorna tadorna*).
Spot-billed Duck (*Anas poecilorhynchus*).
Cinnamon Teal (*A. cyanoptera*).
Garganey (*A. querquedula*).
Chestnut-breasted Teal (*A. castanea*).
Common Teal (*A. crecca*).
Falcated Teal (*A. falcata*).
Versicolor Teal (*A. versicolor*).
Chilian Pintail (*A. spinicauda*).
Common Pintail (*A. acuta*).
Bahama Pintail (*A. bahamensis*).
Common Wigeon (*Mareca penelope*).
Chiloe Wigeon (*M. sibilatrix*).
Shoveler (*Spatula clypeata*).

Carolina Duck (*Aix sponsa*).
Mandarin Duck (*Dendronessa galericulata*).
Rosy-billed Duck (*Metopiana peposaca*).
Common Pochard (*Nyroca ferina*).
American Red-headed Pochard (*N. americana*).
Tufted Duck (*N. fuligula*).
Barrow's Goldeneye (*Bucephala islandica*).

Escapes from one or other of these centres are constantly turning up on reservoirs and sewage farms near London ; those reported within recent years include the sarus crane (*Megalornis antigone*), black swan, ruddy sheld-duck (*Casarca ferruginea*), American wigeon (*Anas americana*), Bahama pintail (*Anas bahamensis*), red-crested pochard (*Netta rufina*) and ferruginous duck (*Aythya nyroca*) ; the last-named are bred in St. James's Park and full-winged young are not infrequently reported from other waters in the autumn. Three species of water-fowl which have escaped from captivity in the London area have succeeded in establishing themselves as resident breeding species, the Canada goose, the gadwall and the tufted duck. The two last of these are native British breeding birds in their own right.

The Canada goose has been domesticated as an ornamental water-fowl in Britain since the seventeenth century, and was stated by Latham in 1785 to breed freely at many country seats. In the London area at present its chief stronghold is at Gatton Park near Reigate, Surrey, where several pairs breed, and upwards of 200 are sometimes seen in winter, attracted by the food put down for the mallard which are preserved there. In the summer the geese scatter over the country-side and breed at numerous small ponds and lakes in the Weald. Occasional pairs have nested in other parts of the London area, once as near to the centre as Walthamstow Reservoirs.

The gadwall has established itself at Barn Elms Reservoirs, where it has been seen regularly since 1933 and has bred since 1936. The original colonists were almost certainly the full-winged young of pinioned birds from St. James's Park. Since the war observation at Barn Elms has not been possible, but gadwalls can frequently be seen on the river nearby, so there is every reason to suppose that they are still breeding at Barn Elms. As more gadwalls are to be seen in winter than at other seasons, it is also supposed that they may be breeding in other parts of the London area, though the only definite breeding record elsewhere is for Beddington, Surrey, in 1938. A mallard-gadwall hybrid was seen at Barn Elms in March, 1939, by Dr. G. Carmichael Low. South of the Border, the only other part of the British Isles where the gadwall breeds is on the breckland meres of East Anglia, where it was also introduced, about a hundred years ago.

The tufted duck has spread very widely over the British Isles as a breeding species in the past fifty years, so it would be incorrect to attribute its present status as a breeder in some numbers in the London area entirely to the escape of full-winged young from the Central Parks. There is no doubt, however, that this is a factor, though nobody has traced out the story in detail. In 1913 full-winged tufted ducks were nesting freely in the London parks, as they do to-day, and they have also bred on waters as near the centre as Barn Elms and Stoke Newington Reservoirs and Clissold and Finsbury Parks. On the Highgate Ponds, on the other hand, they are still only winter visitors, arriving in October and departing in April or May.

Among the most popular of the ornamental waterfowl to be seen in London are the pelicans (Plate 26b) and cormorants (Plate XXXIa) of St. James's Park. The pelicans were there as long ago as the reign of Charles II, when they were seen by Evelyn. The cormorants are of more recent origin, having been brought from the Megstone Rock, most northerly of the Outer Farnes, in 1888. Wild cormorants, and also shags, visit the London area almost annually, and are most often seen by the Thames. Once a cormorant was seen perching on the topmost point of St. Paul's Cathedral.

The swans of the Thames, though living a completely feral existence, are still claimed as the private property of the Crown, and the Vintners' and Dyers' Companies, who mark all the young swans each year to prove their claims. At the swan-upping in 1943 a total of 464 birds were counted from the estuary up to Windsor ; of these 283 belonged to the Crown, 105 to the Vintners and 76 to the Dyers. Six of the Vintners' birds were black swans, so that we here have a medieval custom extended to cover birds from Australia, a part of the world that was unknown when the custom began. In hard weather large numbers of swans gather between Hammersmith Bridge and Chiswick Eyot (Plate 20a), and odd pairs may be seen throughout the Thames's course through London, even in the most wharf-lined areas, as well as along the Lea (Plate 24b) and on many ponds and artificial lakes, such as the Hampstead and Highgate Ponds. Within the

London area, swans are by no means sedentary, and move about quite frequently from one pond or stretch of the river to another.

It is but a step from the keeping of animals as pets, either in or out of doors, to the treating of genuinely wild creatures as friends. One of the commonest sights of London is to see people feeding the birds—pigeons in Trafalgar Square or St. Paul's Churchyard (Plate 7), sparrows in the parks (Plate XVa), or gulls on the Embankment (Plates 21a, XXI). Hudson dated the practice of feeding the gulls to the severe winter of 1892-93, when

> " every day for a period of three to four weeks hundreds of working men and boys would take advantage of the free hour at dinner time to visit the bridges and embankments, and give the scraps left from their meal to the birds."

Equally familiar is the sight of food put out for the birds in suburban gardens—coconuts and lumps of fat for the tits (Plate XIVb), and crumbs for the sparrows, starlings, blackbirds and thrushes. In severe weather, when snow and frost seal up the natural sources of food, the scattering of scraps saves many thousands of lives of birds in the London area. In the cold spell of December, 1938, for instance, greenfinches, skylarks, and even a brambling were among the unusual visitors to London gardens, as well as increased numbers of the regulars, such as sparrows, starlings, chaffinches, tits, blackbirds, thrushes and robins. The brambling came to a garden in St. John's Wood and fed on shelled pea-nuts. Again in the cold spell of the 1939-40 winter, many unusual birds came to be fed in gardens round London. They included another brambling, skylarks, a meadow pipit, fieldfares, redwings and both common and black-headed gulls. At Cheshunt, Herts., fieldfares and redwings came to a garden during the cold spell, but would not accept food.

Altogether, the feeding of birds is quite an important factor with certain species in the London area. It seems unlikely that there would be quite so many pigeons in the streets, ducks (Plate 21b) and woodpigeons in the parks, gulls on the river, or tits in the gardens, if they were not actively encouraged by being fed. The common gulls on the Embankment are even developing parasitic habits on the black-heads. They will never come and take food directly from a man, but always prowl about on the fringe of one of the screaming flocks of black-heads, and pursue any that emerge successfully with a scrap, adopting the same tactics as skuas. There does not seem to be a reference to this predatory habit of the common gull in any standard work.

Active protection of birds has been in force in the London area for about fifty years. At present the County of London is governed in this respect by the Wild Birds Protection (County of London) Order, 1909, and Middlesex by a similar Order dated 1935. The Orders are very comprehensive, and include many birds, such as the honey-buzzard, hobby, osprey and bearded tit,[1] that have not been seen in London for very many years, and are not likely to be seen again for as many more. The eggs of many species are also protected.

The chief beneficiaries of bird protection in the London area have probably been the goldfinch and the great crested grebe, which used to be much harried to provide cage-birds and ornamentation for ladies' hats respectively. The goldfinch has undoubtedly greatly increased as a breeding species in the past fifty years, and this has probably been due partly to protection and partly to a decline in the popularity of keeping birds in cages. The great crested grebe has also benefited from a change in fashion—millinery styles happily rarely last long enough to exterminate a species—with the result that a bird which was counted among the rarest British aquatic breeding birds in the nineteenth century has in the twentieth colonised almost every considerable water in the London area. Hudson, in 1898, hardly mentioned this grebe, and it was not till the following year that a pair first appeared and nested on the Pen Ponds in Richmond Park, which ever since have remained a favourite haunt. Two pairs nested in the Park in 1940, the last year for which we have records. By 1931 as many as sixty-eight pairs of these handsome waterfowl were breeding on the score of different waters in the London area, the nearest to the centre being the Walthamstow Reservoirs and Eagle Pond, Snaresbrook, in Essex, the Brent and Stoke Newington Reservoirs and Gunnersbury and Osterley Parks in Middlesex, and South Norwood, Wimbledon and Richmond Parks in Surrey.

More important than the formal legal protection of birds has been the provision of sanctuaries, either deliberately, or indirectly by the creation of public parks. The Royal Parks in and near London—Green, Hyde, Regent's and St. James's Parks and Kensington Gardens in the centre, and Bushy, Greenwich, Hampton Court and Richmond

[1] *Pernis apivorus, Falco subbuteo, Pandion haliætus, Panurus biarmicus.*

Parks and Kew Gardens in the suburbs—have all been made official bird sanctuaries, and a committee set up by the Office (now Ministry) of Works publishes an annual report on birds seen in them. The London County Council also protects the birds in its important open spaces such as Clissold Park, Hampstead Heath and Ken Wood in the north, and Brockwell Park, Peckham Rye, Clapham, Streatham, Tooting Bec, and Wandsworth Commons in the south. There are few private bird sanctuaries or nature reserves of any size near London, the most notable being the Selborne Society's at Perivale Wood in the lower Brent valley. The National Trust maintains bird sanctuaries at Selsdon near Croydon and at Watermeads near Mitcham.

The City Corporation has a remarkably good record in the purchase of open spaces in Greater London for the enjoyment of the citizens. It owns Epping Forest and Wanstead Park in Essex, Highgate and Queen's Woods in Middlesex, Farthing Downs near Coulsdon, Surrey, and Burnham Beeches, Bucks., as well as some smaller areas. The part played by the City Corporation in the preservation of Epping Forest as an open space is one of the classic cases in the history of commons preservation in the nineteenth century, but they were only enabled to do so by the determined vigilance of a Loughton labourer, Thomas Willingale, who in 1866 persisted in maintaining his right to lop hornbeams in the Forest (Plate 5) under a grant made by Queen Elizabeth to the parishioners. The neighbouring Hainault Forest (Plate 27) was less fortunate ; it belonged to the Crown, which in 1851 disafforested the bulk of it, and converted the land to farms. Fifty years later, ironically enough, the London County Council bought one of these farms for a public open space.

Within the past ten years the London County Council has co-operated actively with other county councils in the London area to preserve the remaining fragments of London's countryside under the " Green Belt " scheme.

The flora and fauna of the parks and open spaces in the partly built-up areas, and especially the recent increase in certain species, such as the woodpigeon and moorhen, have already been discussed in Chapter 9.

The relationships between man and plants are roughly similar to those between man and animals. Some plants are kept in the home, others are cultivated out of doors in gardens, and yet others, which grow wild, are either ravaged or protected, as the case may be. It is

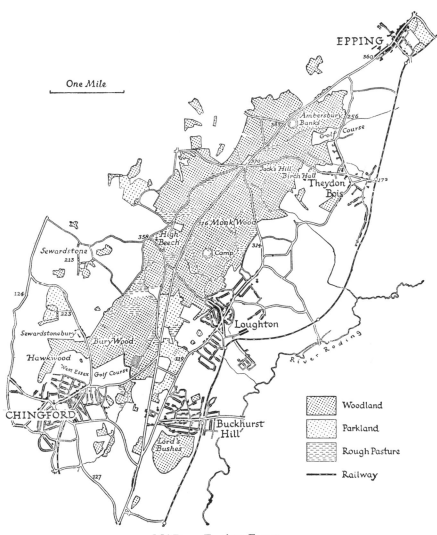

One Mile

EPPING

360

Ambersbury
Banks 256

Golf Course

370

Jack's Hill

Birch Hall

Theydon
Bois

172

316 Monk Wood

314

359

High
Beech

Camp

Sewardstone
213

124

223

Sewardstonebury

Bury Wood

Hawkwood

West Essex Golf Course

119

River Roding

Loughton

CHINGFORD

Lord's
Bushes

Buckhurst
Hill

127

| Woodland |
| Parkland |
| Rough Pasture |
| —— Railway |

MAP 9   Epping Forest.
(Based on the 1″ Ordnance Survey Map, 1942)
By permission of H.M. Stationery Office

223

not customary to regard the plants kept in pots in the home as pets, yet in a way the aspidistras (Plate XVIIIb), hyacinths, cyclamens and ferns to be found in so many London homes are the plant equivalents of cats and dogs. Window-boxes, usually containing geraniums, marguerite daisies and other popular garden plants, often brighten the window-sills of the dingiest tenements, and may be regarded as the plant equivalents of the canaries and other birds whose cages hang outside the same windows.

A halfway house between the keeping of plants in pots and cut flowers in vases indoors, and the growing of flowers in gardens, is the hothouse, greenhouse or conservatory that is attached to many of the more prosperous houses in London. The numerous hothouses of Kew Gardens have been the unintentional cause of many strange imported creatures surviving in England. Among them was a tree-frog from the West Indies (*Hylodes martinicensis*), which flourished for some years in the 1890's, hiding during the day among plant pots and orchid baskets and emerging at dusk to utter a shrill whistling call, not unlike the piping of a nestling bird. Several exotic wood-lice, and a fresh-water snail (*Physa acuta*) that is common in pools in the West Indies and the south of France, have also occurred at Kew.

The importance of gardens as a habitat for both animals and plants has been discussed in Chapter 9. Here it is only necessary to point out that gardening is one of the most popular pastimes of Londoners, and that in their zeal for stocking their gardens and rockeries they have uprooted some of the more striking plants of the London area almost to the point of extinction. Both primroses (*Primula vulgaris*) and foxgloves (*Digitalis purpurea*) have suffered heavily from the amateur gardener's trowel, and it is doubtful if any now grow wild within ten miles of Charing Cross, unless in preserved grounds, while few can be found even within fifteen miles. Robbins in 1916 attributed the extreme scarcity of these and other common woodland plants in Epping Forest to its proximity of London and the large number of visitors. Ferns have suffered particularly from these depredations, so that only a few species, notably bracken (*Pteridium aquilinum*), male fern (*Dryopteris filix-mas*) and broad shieldfern (*D. spinulosum*), are at all common within the twenty-mile radius. L. G. Payne quotes a case from the last century when the royal fern (*Osmunda regalis*) was bought from a tramp in Fleet Street. To-day this magnificent fern is quite extinct within twenty miles of London, and Salmon

PLATE 39

Berkeley Square plane trees   L. DUDLEY STAMP

Glasshouses in the Lea Valley ERIC HOSKING

Cress-beds at Fetcham, Surrey ERIC HOSKING

gives fifty-five stations for it in Surrey, in only four of which was it seen between 1900 and 1931.

The fine displays of flowering plants, such as crocuses (Plate 28b), daffodils (Plate 28a) and tulips (Plate 29) in the spring, and dahlias and chrysanthemums in the summer, in the London parks, together with the many fine trees and shrubs, such as horse-chestnuts (Plates 30, V), magnolias (Plate 19), ginkgos (Plate 11), almond and other fruit trees (Plates 31a and 32), azaleas (Plate 33) and forsythia (Plate 34), to be found in them, are additional evidence of the close senti-mental ties between Londoners and the plant world. One of the most interesting of these parks is the Chelsea Physic Garden (Plate IX), founded by the Apothecaries' Society in 1673 on the present site, which was given to them fifty years later by Sir Hans Sloane, as " a physic garden, so that apprentices and others may better distinguish good and useful plants from those that bear resemblance to them and yet are hurtful." The garden, which can be visited only on application to the curator, is still used for scientific purposes, especially in connec-tion with the work of the Imperial College of Science.

The gardens of the Royal Botanic Society in Regent's Park have recently been thrown open to the public, and are now known as Queen Mary's Garden (Plate 31b). They are noted for their fine display of roses. The famous Royal Botanic Gardens at Kew, Surrey, were founded by the Princess Dowager of Wales, mother of George III, in 1760, and were presented to the Nation by Queen Victoria in 1841. Their directorship is one of the most important posts that can be bestowed on a British botanist to-day. The importance of Kew Gardens in the natural history of London has been evident throughout this book, which has made frequent reference to its displays of flowers and trees, the alien species which either inhabit it or have spread from it, and its importance as a bird sanctuary.

Gardeners are not the only vandals with whom wild plants near London have to contend. Hikers and others pick large quantities of flowers each year for indoor decoration, and at Christmastide the hollies and other evergreens are regularly raided for the same purpose. It is a familiar sight in May on the roads leading into London to see cyclists with their carriers stacked with bluebells (*Scilla non-scripta*) (Plate 35), most of which invariably wilt away before their optimistic ravishers reach home. The movement for the protection of plants has achieved much smaller success than that for the protection of birds.

On the more popular open spaces, as mentioned in Chapter 9, the sheer weight of trampling feet has driven out practically all flowering plants, except for such tough species as the yarrow (*Achillea millefolium*) and lowly ones like the Dutch clover (*Trifolium repens*). This has also happened in one or two public woods. Highgate Wood, for instance, now has no undergrowth, except for a few scanty brambles (*Rubus fruticosus*), and its floor is as bare of vegetation as any fir plantation, though in fact it was originally a dry oakwood, as can be seen from the small patch which has been fenced round as a bird sanctuary (Plate 36.) Thirty years ago Mr. Charles Nicholson, addressing the London Natural History Society, bewailed the fact that

> " Highgate Woods, which thirty years ago were a blaze of colour in spring, are now so intersected with cinder and asphalt paths that their beauty has been almost destroyed. Bluebells are practically non-existent, and the thousands of wood anemones are now represented by a few miserable clumps of leaves here and there. The better drainage of the woods has destroyed numerous plants, and several fine clumps of such plants as *Carex pendula* and *Carex vesicaria* have been lost. . . . The acquisition of Hampstead Heath for public use has now practically extinguished its interest as a botanical area. . . ."

The other side of the picture, however, is the great and increasing interest in the wild animals and plants of the London area that is displayed by Londoners to-day. On Sundays in the spring and summer, and also in more inclement seasons, the trains going out to the country districts are filled with Londoners bent on getting a few hours in the fresh air. Few are so insensitive to their surroundings as to care nothing at all for the natural components of the rural environment, and many show an active interest and enjoyment in identifying and appreciating the numerous animal and plant species that can be seen on a walk in such favourite spots as Epping Forest, Burnham Beeches, Box Hill, the Chess valley and Richmond Park. Increasing numbers take the active step of joining a body, such as the London Natural History Society or the British Empire Naturalists' Association, which caters for both novices and *cognoscenti* by organising rambles and excursions in places where interesting species are likely to be met with. A list of such societies will be found in Appendix E.

References for Chapter 17

Brogan (1943), Collenette (1937), Committee on Bird Sanctuaries in Royal Parks (England) (1929-39), Drewitt (1928), Eversley (1910), Finn (1905), Fitter (1939, 1941a), Fitter & Homes (1939), Günther (1906), Hampstead Scientific Society (1913), Harrisson & Hollom (1932), Harting (1866), Horn (1923), Hudson (1898), London Natural History Society (1900-44, 1937-44), Nicholson (1916), Payne (1942), Pigott (1902), Ritchie (1920), Robbins (1916), Salmon (1931), Stubbs (1917a), Zoological Society of London (1939).

THE INFLUENCE OF THE WAR

NOT since London ceased to be subject to recurrent fires, due to the carelessness of its citizens combined with the combustibility of their building materials, has the urban area been subjected to such a cataclysm as the " blitz," which raged from September, 1940, to May, 1941, and was followed three years later by the " fly-blitz." The effect of what is now euphemistically known as " the original blitz " has been to produce extensive areas of open ground throughout the most heavily built-up parts of London, and this has naturally had a profound influence on the fauna, and more especially on the flora of the area. Apart from the aftermaths of previous fires, there is more open ground to-day within a mile of St. Paul's Cathedral than there has been since the early Middle Ages. There is even cultivation, pigs being kept and vegetables grown on an allotment near Cripplegate Church, which shows that even this land, built over for seven or eight centuries, still retains its fertility. It is sometimes forgotten that much of Greater London stands on some of the most fertile soil of the British Isles ; it just happens that under the present order of society it is more profitable to grow factories than fruit or vegetables.

Comparatively few animals and plants in the London area have fallen direct victims to the *Luftwaffe*. Every time a bomb falls on open land, of course, so many plants are physically disintegrated, and trees suffer from the effects of blast, but in no case can the status of any plant have been affected. An interesting correspondence in *The Times* gave several examples of trees leafing or flowering again in the autumn after having been stripped by the blast from flying bombs. A horse-chestnut tree in Camberwell that was stripped of almost all its leaves in July was in full bloom again in September, and at Reigate, Surrey, lilac bloomed again and a creeper came into full leaf after its first crop had all been blown into the house.

Animals, being more mobile, have probably suffered less than plants from the bombing, apart from the innumerable invertebrates that have been blown sky-high along with the soil and turf. One of the few actual casualties that have been reported was a mistle-thrush that was killed by blast at Watford, being found dead thirty yards away from the crater. During the height of the " original blitz " the author received a request from a biologist in the north of England for any corpses of starlings that might be killed during the raids ; unfortunately for the biologist, though fortunately for the starlings, which continued to roost in London throughout the bombing, no corpses were ever reported. On at least one occasion bombs that fell in the Thames killed some fish and a number of crows and gulls were seen feeding on them.

Animals, and especially birds, are liable to take fright at the noises of modern aerial warfare, but the evidence, both in this war and the last, on this point is conflicting. Any one who has stood in Trafalgar Square for five minutes will have noticed that the pigeons take fright and fly round in a body at any loud noise, such as a back-fire. Much of the fright that has been reported in birds during air raids is an extension of this simple protective device ; in Nature loud noises usually bode no good. When the bells of St. Paul's were rung for the victory of the tanks at Cambrai in November, 1917, and again when they were rung for the final victory a year later, the pigeons took fright even at this sound which had once been so familiar to them that they ignored it. Several examples of the alarm of birds during air raids in both wars have been recorded ; during the last war the sparrows and owls in Kensington were very restless for an hour or so before the earlier raids, but afterwards became accustomed to them ; one of the Kensington sparrows is said to have fainted during a raid ; Sir Hugh Gladstone observed the London pigeons very much alarmed during the daylight raid on London on July 7, 1917, " whirling about in flocks in a thoroughly scared manner " ; in 1940 a heron flew screaming over a garden in St. John's Wood after the explosion of a delayed action bomb nearby, and on a similar occasion eight frightened crows also flew over. One result of the " fly blitz " of 1944 was to drive many of the woodpigeons from the London parks.

That birds are not always frightened by strange loud noises, and that the effect of any fright is certainly short-lived, is suggested by many other records. Nightingales are well known for their indifference to gunfire, which indeed they often seem to regard as a particularly vigorous rival nightingale trying to muscle in on their territory ; during a raid in May, 1918, for instance, one was singing loudly in a London suburb during heavy gunfire and bomb explosions. Again, in the daylight raid of July 7, 1917, when Sir Hugh Gladstone saw the frightened pigeons, Mr. Charles Dixon saw pigeons, sparrows and starlings moving quite unconcernedly about the roads barely a hundred yards from where shrapnel was bursting, while thrushes sang on and off through-out the raid period. Whether the fact that a barn-owl was screeching at Norbury during an anti-aircraft barrage in 1940 was due to fear, rage or just unconcern while the owl went about its normal business, is anybody's guess. Certainly the fact that a swift circled around during a fierce air battle over the Roding valley in 1940 may be imputed to relative unconcern in the swift. Starlings, as we have seen, continued to roost in London throughout the noisy and death-fraught winter of 1940-41, nor have they substantially altered their habits since. The great heronry at Walthamstow was hardly affected at all by the hail of bombs that fell on all sides of it, though none actually hit the heronry island. There was in fact an increase of three occupied nests in 1941 compared with 1940, in contrast to the decrease in heron breeding population in the rest of the Thames valley and over the whole country.

A more serious type of direct war interference with animal and plant life has been the havoc necessarily wrought in places set apart for military training. Modern warfare consorts ill with nature con-servation, and an area that has been used for battle training soon contains more mud than grass. Fortunately, the London area is too built-up for there to be much open space that can be mutilated in this way, but from one area at least, Richmond Park, comes a story of fewer breeding birds. Fewer redstarts and whitethroats nested in 1940 than in 1939, and a cock whinchat disappeared after being alarmed by soldiers training. The well-known heronry also suffered severely, its numbers falling from sixty-one breeding pairs in 1939 to forty in 1942, of which nine eventually deserted ; but this was partly due to the felling of trees. Nothing has yet been reported in this war to equal the resourcefulness of a colony of sand-martins that nested in some trenches dug for military training at Gidea Park, Essex, in 1917.

Foremost among the effects of the war on the natural history of London has been the creation of extensive areas of open waste ground all over the built-up area, and especially in the City and the eastern boroughs of the County of London (Plates Xb, XXXII). The flora of these areas has been carefully studied by Dr. E. J. Salisbury, F.R.S., Director of the Royal Botanic Gardens at Kew, and by Mr. J. E. Lousley, on whose papers the following paragraphs are mainly based. Dr. Salisbury's complete list is given in Appendix F.

Generally speaking, there are more plant species on the western than on the eastern waste sites in London. This is due partly to the direction of the prevailing wind, and partly to the fact that there are more gardens in West London, from which weed and other seeds can spread. Since over large areas of the inner built-up core of London there was formerly no bare ground at all where plants other than mosses and algæ could grow, whatever has sprung up since 1940 must either have originated in wind-borne seeds, or else have arrived on the feet or clothing of men and other animals, including the nose-bags of horses. It will be recollected that the colonisation of five waste building sites in London between 1901 and 1910 was attributed by Shenstone to the same causes. (See Chapter 8.)

Dr. Salisbury observes that whereas the London rocket (*Sisymbrium irio*) grew abundantly among the ruins after the Great Fire of 1666,[1] he failed to find a single specimen on any of the sites he examined after the Great Fire of 1940, nor has any other reliable observer reported it. Its place, as we have seen, has been taken by the rose-bay willow-herb (*Epilobium angustifolium*) (Plate 37), which is at present probably the commonest plant in Central London, and has brought with it many elephant hawk-moths (*Chærocampa elpenor*), whose larvæ feed on its leaves. In 1869 the rose-bay was regarded as a rare denizen of gravelly banks and woods, and was allowed by Trimen and Dyer only eight stations in the whole of Middlesex, including Ken Wood and Paddington Cemetery. It seems to owe its latter-day prominence to a liking for plenty of light, which of course it gets on the open blitzed sites, combined with a tolerance for soil that has been subjected to heat, which enables it to get a firm hold before its competitors. Moreover, a single young plant is capable of producing some 80,000 seeds in one season, each one of which is adapted to float on the slightest breeze, so that it is not surprising that Dr. Salisbury found rose-bay on 90 per cent of his sites.

Next in importance come three members of the groundsel family,

---

[1] In the *Catalogus Angliae* (1670) Ray recorded the London rocket in various places on mounds of earth between the City and Kensington ; after the Fire it came up in the greatest plenty in 1667 and 1668 within the walls on the " rubbish-heaps around St. Paul's Cathedral."

the common groundsel (*Senecio vulgaris*), known to every gardener, the Oxford ragwort (*S. squalidus*) and the sticky groundsel (*S. viscosus*). The advantages of the common groundsel in the struggle for existence include an output of 1100 potential seedlings per plant, glutinous seeds that can be dispersed on the feet of men and birds as well as by the wind, the possibility of several generations ripening their seeds in a single season, and the capacity of the seeds to germinate at almost any time of the year when the temperature is favourable. The Oxford ragwort (Plate XXXII), like other things bearing the name of that city, is an alien, a native of Sicily, where it frequents volcanic ash, so that, as Dr. Salisbury remarks, it may well find the site of a burnt-out building a congenial habitat. It was first recorded in England in 1794 at Oxford, where it had probably escaped from the Botanical Gardens, and did not arrive in London till 1867. The sticky groundsel is one of those plants about whose status botanists are always disputing ; it may be native in a few localities, or it may be adventive everywhere In London, at any rate, it is if not an alien a provincial, for it was first noticed near the Caledonian Hospital in Copenhagen Fields in 1838, and was described by Trimen and Dyer in 1869 as an " accidental waif." In Dr. Salisbury's survey it was found on 45 per cent of the sites, compared with 56 per cent which held its Oxonian relative.

The coltsfoot (*Tussilago farfara*), a very common plant of waste ground all round London on building sites awaiting development, railway embankments, roadsides and similar places, is, as one might expect, also found on many blitzed sites. (Plate 38.) Its habit of flowering in February and March enables its seeds to ripen and the seedlings to get well established before competitors arrive.

Another invader that has muscled in on the blitz is the Canadian fleabane (*Erigeron canadensis*), which was first recorded in London in 1690, when Ray stated on the authority of Tancred Robinson that it occurred freely but was certainly not indigenous. An American writer, Marsh, has stated that this plant originated in Europe from a seed which fell from a stuffed bird, but no corroborative evidence has ever been produced. The Canadian fleabane was abundant on the site of the Exhibition of 1862 at South Kensington, and in 1877 was noted as frequent on waste ground and especially on railway embankments near London. It is present on about 40 per cent of the London bombed sites, and since each plant can produce some 120,000 fruits, it would seem to be merely a matter of time before it appears on them all.

Other composites found by Dr. Salisbury include the sow-thistle (*Sonchus oleraceus*), dandelion (*Taraxacum officinale*) and the South American alien *Galinsoga parviflora*. The last-named, found on 14 per cent of the bombed sites, is a native of Peru that was first recorded in England as a garden weed at Richmond, Surrey, in 1860. It had escaped from Kew Gardens.

Among the grasses recorded by Dr. Salisbury as occurring fairly commonly on the bombed sites are the common and smooth meadow grasses (*Poa annua, Poa pratensis*), rye-grass (*Lolium perenne*), marsh bent-grass (*Agrostis alba*), Yorkshire fog (*Holcus lanatus*), cocks-foot grass (*Dactylis glomerata*), oats and wheat. The presence of the two last-named suggests horses' nose-bags as the distributive agent, and the same source is probably also responsible for the frequent presence of the two common clovers, the red (*Trifolium pratense*) and the white or Dutch (*T. repens*). The red clover was found on a quarter of the sites examined.

To pedestrians' boots must be attributed the occurrence of such common weeds as the chickweed (*Stellaria media*) and the plantain (*Plantago major*), both found most commonly in areas to which pedestrians have easy access.

The total number of vascular plants recorded by Dr. Salisbury from sites which might reasonably be assumed to have been previously devoid of vegetation (i.e., excluding the sites of former gardens) was 126. Mr. Lousley, who examined twenty-one sites in the first two years after the blitz (1941-42), found only twenty-seven species, showing that it takes time for the great variety now to be found to colonise the sites. The only plants found by Mr. Lousley on more than two of his sites were rose-bay, Canadian fleabane, coltsfoot, Oxford ragwort, common and sticky groundsels, sow-thistle and common meadow-grass. These are the pioneer colonists of the bombed sites.

Dr. Salisbury found that wind carriage was by far the most important agent for introducing plants on to the bombed sites, 30 per cent of all the species having probably arrived in this manner. Of the species found on over 25 per cent of the sites, eight were wind-dispersed and only four dispersed by other methods. Non-windborne composites, such as the yarrow (*Achillea millefolium*), mugwort (*Artemisia vulgaris*), scentless mayweed (*Matricaria inodora*), nipplewort (*Lapsana communis*) and pineapple weed (*M. suaveolens*), were found on only a small proportion of the sites. Moreover, the only garden plant found, the purple buddleia (*Buddleia davidii*), is a windborne one.

Of the two commonest tree saplings, the sallow (*Salix caprea*) has windborne seeds, and was found on 16 per cent of the sites, but the elder (*Sambucus nigra*) was probably bird-borne. Birds were probably also responsible for the occasional presence of the woody and black nightshades (*Solanum dulcamara, S. nigrum*), while the not infrequent presence of tomato seedlings was most likely due to the refuse from office-workers' lunches. Altogether, Dr. Salisbury estimated that 20 per cent of the special were present on account of avian and 14 per cent on account of human dispersal.

Mr. C. P. Castell has given an interesting account of the vegetation of a small pond in the crater of a bomb that fell on Bookham Common, Surrey, in the winter of 1940-41. No plant life was visible in June, 1942, but the branched bur-reed (*Sparganium ramosum*) and a species of rush (*Juncus*) were present a year later. By 1944 the plant community of this crater pond comprised the branched bur-reed, which was dominant, the water plantain (*Alisma plantago-aquatica*), and four species of rush (*J. articulatus, J. conglomeratus, J. effusus* and *J. inflexus*).

On the animal side the major development that can be attributed at any rate in part to the blitz is the remarkable spread of the black redstart as a breeding species (see Chapter 8). Since 1942 all the nesting sites of this species that have been located in London have been in crannies created by the bombing. House-sparrows, too, have been presented with many suitable new nesting sites, and an observer at Watford in 1940 found that they were quick to explore the new nooks and crannies in a bombed house. This may, however, have been more in the nature of a search for food than for nesting sites, as the same observer noted that they showed a marked partiality for new putty, possibly because of the linseed oil it contained.

Wheatears have sometimes mistaken the brick-strewn surface of bombed sites for the rocky and stony places they often frequent in their breeding quarters. One was seen flitting about the ruins of a devastated area in Stepney in October, 1940, and another was feeding on the blitzed area near Cripplegate churchyard in September, 1942. A wheatear was also seen on the roof of a block of flats in Kensington in September, 1940. Another interesting development in connection with the bombed sites is that the growth of weeds on some of them has become so dense that warblers, such as whitethroats and willow-warblers, have been seen skulking in the undergrowth on migration right in the heart of London.

One of the minor effects of war is the draining of ponds for military reasons. In the 1914 war the lake in St. James's Park was drained, but we have been spared this deprivation in the present war. When the Pen Ponds in Richmond Park were drained in 1940, a large number of fish were removed, and in many cases transferred to other ponds in the neighbourhood. They included 175 carp, 320 bream, 250 pike, 300 eels, 30 perch and 20 dace. It was estimated that about half the fish were rescued ; many of the larger ones were out of reach in the mud, and perished. The largest pike weighed 11 lb., the largest carp 8½ lb., the largest bream 8 lb., and the largest eel 3½ lb. The older fish were a very mixed lot ; many were in bad condition, odorous, verminous and unfit for human consumption. The younger ones were healthier, and were mainly transferred to other ponds. All the pike and eels were destroyed, those fit for human consumption being eaten. While the ponds were being drained, three bombs fell in them, and a number of fish were killed.

When the ponds were drained, large numbers of freshwater mussels, measuring up to six inches long, were found on the bottom, but black-headed gulls soon cleared the lot. In the following spring plants began to appear, the first species noted being a large patch of celery-leaved crowfoot (*Ranunculus sceleratus*). By the time the mud had dried out in the summer, the whole surface of the ponds had become a three-foot high jungle of rushes and weeds. Mr. D. A. Rawlence recorded the following plants there in August, 1941 :

Lesser spearwort (*Ranunculus flammula*).
Celery-leaved crowfoot (*R. sceleratus*).
Lesser celandine (*R. ficaria*).
Marsh watercress (*Nasturtium palustre*).
Codlins and cream (*Epilobium hirsutum*).
Broad-leaved willow-herb (*E. montanum*).
Bur-marigold (*Bidens*).
Wood-groundsel (*Senecio sylvaticus*).
Coltsfoot (*Tussilago farfara*).
Gipsywort (*Lycopus europaeus*)
Water mint (*Mentha aquatica*).
Common Skullcap (*Scutellaria galericulata*).

Dock (*Rumex*).
Knotgrass (*Polygonum aviculare*).
Amphibious persicaria (*P. amphibium*).
Red persicaria (*P. persicaria*).
Pale persicaria (*P. lapathifolium*).
Waterpepper (*P. hydropiper*).
Lesser reed-mace (*Typha angustifolia*).

All these were growing amidst a forest of common and hard rush (*Juncus conglomeratus, J. inflexus*), with seedling birches and willows springing up everywhere.

An interesting phenomenon allied to the draining of ponds occurred when Teddington lock was bombed, so that the river above it became tidal for a time. All the mallard from some distance around used to flight to the newly tidal part of the river every evening just as if it had been far down the estuary.

Whereas certain ponds have disappeared, other smaller ponds in the shape of static water tanks have appeared all over London, and birds have not been slow to make use of them.[1] Grey wagtails are often seen on these tanks, and mallards (Plate XXXIb) have even bred on them. In a letter to *The Field* in February, 1944, Mr L. Jones reported having seen a swan, a tufted duck, and even a greenshank on one static water tank in North London at various times. So far as the mallard are concerned, this is only a development of their pre-war habit of visiting such small ponds in Inner London as the basins in Trafalgar Square, and the pond in New Square, Lincoln's Inn, where a pair nested in both 1930 and 1931, much to the delight of the Benchers. In 1930 also a pair of mallard brought off six ducklings by a small pond in the garden of No. 13 Holland Park, while in 1937 two broods were hatched in the long grass among the tombstones in Brompton Cemetery. It is on occasions like these that mallards, and swans as well, will hold up the traffic while they escort their broods across the road to the nearest water.

On some reservoirs artificial obstructions, such as booms, have appeared as a result of the war, and on one of these at Lonsdale Road Reservoirs a pair of great crested grebes nested for two years in succession.

There has been a big increase in the number of allotments (Plate

[1] Dr. Dudley Stamp informs me that in Sloane Square the pigeons have developed an interesting technique of bathing in static water tanks. He has often watched them take off from the edge and flutter half in the water across to the other side.

13a) in the London area since the war, and many golf courses and other open spaces, such as Bushy Park and Parliament Hill Fields, have been partly ploughed up. This has brought an increase both in the weeds of arable land, and in the birds of the garden association, such as blackbirds and thrushes, which feed either on the seeds of the weeds or on the varied invertebrate life of cultivated ground. A white wagtail that was seen on the allotments near the Albert Memorial in March, 1941, was only the second record for that subspecies in Inner London. The ploughing up of part of Parliament Hill Fields seems to have had little effect on the birds as yet, though it may have been responsible for the presence nearby of six rooks in November, 1942, rooks being rarely seen so near London nowadays, except when flying over. Pied wagtails and chaffinches feed on this ploughland, but both are birds which normally frequent the Heath and its environs. What would be really interesting would be if rooks and lapwings, the commonest feeders on arable land in the country, took to feeding there regularly. Other by-products of the allotment boom that have been reported near London are the decrease of a colony of tree-sparrows near Harrow, and the shooting of woodpigeons and stock-doves in Kensington Gardens to stop their depredations. The shooting of the stock-doves was apparently due to faulty identification, woodpigeons being the sole intended victims, though as stock-doves are almost as fond of peas as woodpigeons, there does not seem to be a case for sparing them on these grounds.

The felling of trees also ranks among the minor effects of the war in the London area. There are so few trees to fell—in the Land Utilisation Survey only 1.5 per cent of the area of London and Middlesex combined was recorded as forest and woodland—that this is not the important factor it is in other parts of the country, where hundreds of acres have been converted from woodland to scrub. The felling of trees has been adduced as a reason for the decrease of the breeding population of herons and tree-sparrows in parts of the London area. The status of woodpeckers is not likely to be so much affected by tree-felling as might appear at first sight, for they prefer old and rotten trees, which are not worth felling in wartime. Woodpeckers are really a sign of bad forestry, but in suburban areas are, of course, often able to thrive on old trees growing in gardens, shrubberies and shelter belts. The great spotted woodpecker, in particular, is not at all uncommon in wooded suburbs like Hampstead and Highgate.

At the beginning of the war many birds kept in captivity were released owing to the evacuation of their owners or the difficulty of feeding them. The Zoo authorities, as already mentioned (Chapter 17) set free a number of birds of prey. Many waterfowl have also been allowed to go free from the large private collections, so that several records of rather tame scaups and pintail that have appeared in Inner London in the past few years are open to suspicion as to their origin. In 1940, for instance, a drake scaup turned up successively on the Round Pond, the Serpentine, the Thames by Cleopatra's Needle, and the lakes in St. James's and Regent's Parks ; when on the Serpentine it foraged for bread with the tufted ducks. It is difficult to believe that this was a truly wild bird, and it may well be the same one that has visited the lake in St. James's Park annually for several winters.

It seems doubtful whether the regulation prohibiting the feeding to birds of any food fit for human consumption has had any appreciable effect on the bird population. Failure to observe it has certainly been widespread. Many people still feed the gulls on the Embankment, and the ducks in the parks, and it is hard to believe that all the bread they use is mouldy. There are, however, no coconuts for the tits, and fewer people can now spare a fat bone, still less a lump of suet for them. With the pigeons in Trafalgar Square, the boot has even been on the other foot, for the local authorities have had to thin them out, the resultant bag being used for pigeon pie.

A consequence of the war that is apt to be overlooked is the fact that there are fewer observers, owing to the absence of many on war service, while many of those who remain are so busy that they have much less time for their hobby. Moreover, many sewage farms and reservoirs are now closed to the public, and can be watched, if at all, only from public rights of way, so that one fruitful source of records is no longer available. The residue of man-hours of bird-watching has had perforce to be concentrated on fewer places, so that in fact some localities, notably the Thames towpath between Hammersmith and Mortlake, have been more intensively watched since the war than before it. During the year 1943, for instance, seventeen observers are known to have paid 200 separate visits to the towpath, mostly in the Hammersmith-Mortlake sector, on 144 different days. The four most popular months for towpath visiting were January, February, October and December, in that order, and the four least popular months, June,

August, July, and May, also in that order. In the four most popular months at least one observer was present on more than half the days in the month—twenty-three out of thirty-one in January. It must be borne in mind that the presence of Iceland gulls on the Thames at Hammersmith in several recent winters has given the locality a special attraction to London bird-watchers. Moreover, it is probable that the number of visits paid by bird-watchers to the causeway of Staines Reservoir, *locus classicus* of rare aquatic birds in Greater London, in each of the five pre-war years was much greater than this.

References for Chapter 18

Castell (1945), Fitter (1941), Gladstone (1919), London Natural History Society (1937-44), Lousley (1944), Macpherson (1929-40), Marsh (1864), Raven (1942) Rawlence (1942), Salisbury (1943), Shenstone (1912), Trimen & Dyer (1869), Willatts (1937).

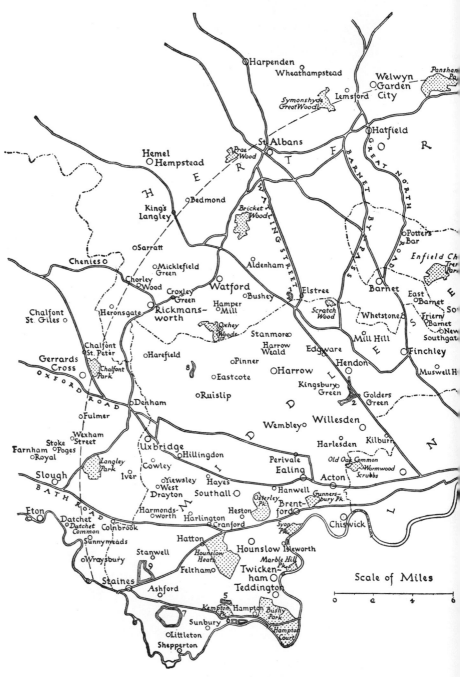

MAP 10 The Greater London Area, North
of the Thames, showing important places.
The radius of 20 miles from St. Paul's

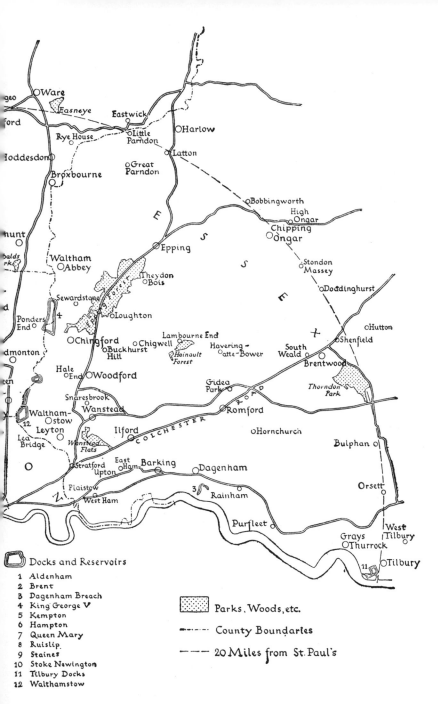

Ware
geo
ford

Easneye

Eastwick

Rye House
Little
Parndon

Harlow

Hoddesdon

Latton

Broxbourne

Great
Parndon

Bobbingworth

High
Ongar

Chipping
Ongar

hunt

balds
rk

Waltham
Abbey

Epping

Theydon
Bois

Stondon
Massey

Doddinghurst

Sewardstone

Ponders
End

4

Loughton

Lambourne End

Hutton

d

Chingford

Chigwell

Havering–
atte-Bower

Shenfield

Buckhurst
Hill

Hainault
Forest

South
Weald

dmonton

Hale
End

Woodford

Gidea
Park

Brentwood

Thorndon
Park

Snaresbrook

Wanstead

Waltham-
stow

Romford

12

Leyton

Ilford

Hornchurch

Lea
Bridge

COLCHESTER

Wanstead
Flats

Bulphan

O

Stratford

East
Upton

Ham

Barking

Dagenham

Orsett

Plaistow

West Ham

3

Rainham

Purfleet

West
Tilbury

Grays
Thurrock

11

Tilbury

Docks and Reservoirs

1 Aldenham
2 Brent
3 Dagenham Breach
4 King George V
5 Kempton
6 Hampton
7 Queen Mary
8 Ruislip
9 Staines
10 Stoke Newington
11 Tilbury Docks
12 Walthamstow

Parks, Woods, etc.

County Boundaries

20 Miles from St. Paul's

Cathedral is indicated. (Based on the ½″
Ordnance Survey Map of Greater London,
1935) *By permission of H.M. Stationery Office*

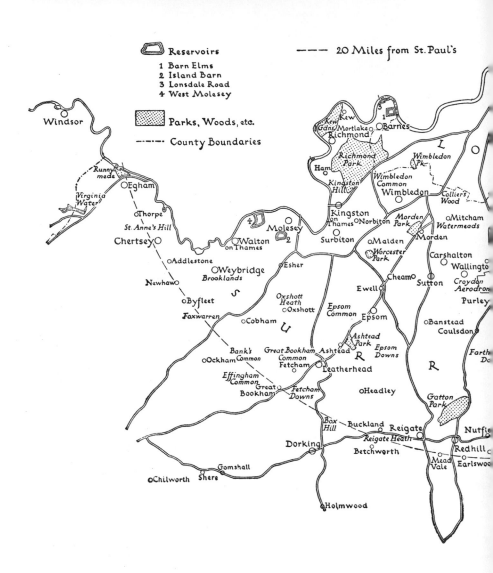

MAP 11 The Greater London Area, South of the Thames, showing important places. The radius of 20 miles from St. Paul's

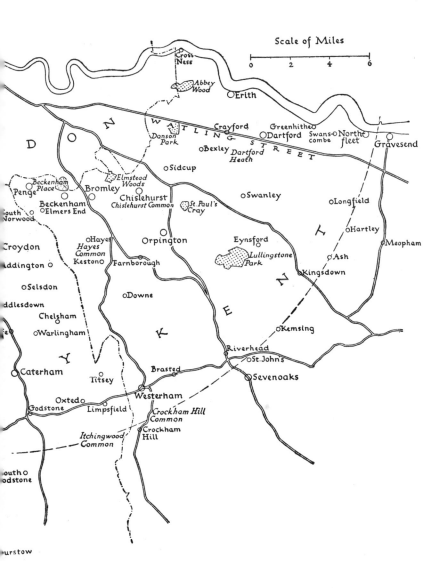

Scale of Miles

Cross Ness

Abbey Wood

Erith

Crayford

Greenhithe

Dartford

Swans-combe

North-fleet

Gravesend

D O N

WATLING STREET

Danson Park

Bexley

Dartford Heath

Sidcup

Beckenham Place

Elmstead Woods

Bromley

Chislehurst

St. Paul's Cray

Swanley

Longfield

Penge

Beckenham

Chislehurst Common

outh Norwood

Elmers End

Hartley

Croydon

Hayes

Hayes Common

Orpington

Eynsford

Meopham

ddington

Keston

Farnborough

Lullingstone Park

Ash

Selsdon

Downe

Kingsdown

ddlesdown

Chelsham

E

Kemsing

Warlingham

K

Caterham

Y

Titsey

Brasted

Riverhead

St. John's

Sevenoaks

Oxted

Westerham

Godstone

Limpsfield

Crockham Hill Common

Itchingwood Common

Crockham Hill

outh odstone

urstow

Cathedral is indicated. (Based on the ½″
Ordnance Survey Map of Greater London,
1935) *By permission of H.M. Stationery Office*

243

A BOTANICAL RAMBLE IN 1629

THE following account of the first recorded botanical ramble in the London area is translated from the Latin of Johnson (1629).

[N.B.—I have marked † plants which I have found on or near the Heath during 1943-45, but the absence of a † should not be taken necessarily to mean that the plant is no longer found there.

R. S. R. F.]

HAMPSTEAD HEATH

At the agreed time in the morning (August 1, 1629) seven of the ten companions[1] met. Buggs, Weale and Wallis were absent, but their places were taken by John Sotherton, John Marriott and Thomas Crosse. We were not deterred by a rainy sky, having prepared for all eventualities ; after overcoming the greater difficulties, we thought shame to give in to a lesser one. Thus, leaving the City, we made our way to Kentish Town, and hastened on from there, but we had not gone far when a heavy shower forced us to seek shelter in Highgate, The storm had scarcely passed when, impatient of delay, we hastened to the wood, where we observed the following plants which we had not seen on the previous [i.e., Kentish] journey[2] :

†False-brome Grass (*Brachypodium sylvaticum* Beauv.).
Wood Spurge (*Euphorbia amygdaloides* L.).
Saw-wort (*Serratula tinctoria* L.).

[1] Jonas Styles, William Broad, John Buggs, Leonard Buckner, Job Weale, Robert Larking, Thomas Wallis, Edward Browne, Edward Browne, Thomas Johnson : they had all toured Kent in July.

[2] Johnson, of course, uses pre-Linnæan names in his plant lists, so it is not always possible to be certain of the species intended. Doubtful identifications are therefore indicated by square brackets. A few names, mainly of ferns and mosses, have defied identification, and have been omitted.

Wood Loose-strife (*Lysimachia nemorum* L.).
Wood Pea (*Lathyrus montanus* (L.) Bernh.).
Alder Buckthorn (*Rhamnus frangula* L.).
†Rowan or Mountain Ash (*Sorbus aucuparia* L.).
†Devil's-bit Scabious (*Scabiosa succisa* L.).
†Wall Hawkweed (*Hieracium murorum* L.).
†Tormentil (*Potentilla erecta* (L.) Hampe).
Dog's Mercury (*Mercurialis perennis* L.).
Square-stalked St. John's Wort (*Hypericum acutum* Moench.).
Aspen (*Populus tremula* L.).
†Birch (*Betula pubescens* Ehrh.).
†Hornbeam (*Carpinus betulus* L.).
Golden-rod (*Solidago virgaurea* L.).
Heath Cudweed (*Gnaphalium sylvaticum* L.).
Creeping Jenny (*Lysimachia nummularia* L.).
†Figwort (*Scrophularia nodosa* L.).
†Angelica (*Angelica sylvestris* L.).
Wood-Sorrel (*Oxalis acetosella* L.).

Leaving the wood, we came on to the Heath itself, and saw the following plants :

Bell-Heather (*Erica cinerea* L.).
Cross-leaved Heath (*E. tetralix* L.).
Whortleberry or Bilberry (*Vaccinium myrtillus* L.).
Lily of the Valley (*Convallaria majalis* L.).
Needle Furze or Petty Whin (*Genista anglica* L.).
Hard Fern (*Blechnum spicant* (L.) With.).
Common Clubmoss (*Lycopodium clavatum* L.).
†Common Skullcap (*Scutellaria galericulata* L.).
Bird's-foot (*Ornithopus perpusillus* L.).
Tutsan (*Hypericum androsæmum* L.).
Round-leaved Sundew (*Drosera rotundifolia* L.).
Common Speedwell (*Veronica officinalis* L.).
Thyme-leaved Speedwell (*V. serpyllifolia* L.).
†Creeping Willow (*Salix repens* L.).
Common Cotton-grass (*Eriophorum angustifolium* Roth.).
Marsh Penny-wort (*Hydrocotyle vulgaris* L.).
Celery-leaved Crowfoot (*Ranunculus sceleratus* L.).
Water Crowfoot [*R. peltatus* Schrank].
†Lesser Spearwort (*R. flammula* L.).
†Foxglove (*Digitalis purpurea* L.).

Whitebeam (*Sorbus aria* (L.) Crantz).
Wild Service-tree (*S. torminalis* (L.) Crantz).

Now leaving the Heath, we refreshed ourselves a little at the township of Hampstead, and then, returning to Kentish Town, dined. In the meadows, field-paths and hedgerows, both going and coming, we saw the following plants :

†Common Orache (*Atriplex hastata* L., and *A. patula* L.).
†Common Vetch (*Vicia angustifolia* L.).
 Hairy Tare (*V. hirsuta* (L.) S. F. Gray).
 Good King Henry (*Chenopodium bonus-henricus* L.).
† Lords and Ladies (*Arum maculatum* L.).
†Herb Robert (*Geranium robertianum* L.).
 Dove's-foot Cranesbill (*G. molle* L.).
 Common Stork's-bill (*Erodium cicutarium* (L.) L'Hérit.).
 Pepper Saxifrage or Sulphur-wort (*Silaus flavescens* Bernh.).
† Lesser Stitchwort (*Stellaria graminea* L.).
 Wintercress (*Barbarea vulgaris* R. Br.).

It is perhaps also worth while giving here the names of a number of plants found in the same localities on May 1, but now partly or wholly withered away, such as :

†Wood Anemone (*Anemone nemorosa* L.).
†Bluebell (*Scilla non-scripta* (L.) Hoffmans. & Link).
 Cowslip (*Primula veris* L.).
 Primrose (*P. vulgaris* Huds.).
†Ladies' Smock or Cuckoo Flower (*Cardamine pratensis* L.).
 Yellow Archangel (*Lamium galeobdolon* (L.) Crantz).
†Bugle (*Ajuga reptans* L.).
†Jack-by-the-Hedge (*Alliaria officinalis* Andrz.).
†Marsh Marigold or Kingcup (*Caltha palustris* L.).
†Lesser Celandine (*Ranunculus ficaria* L.).

Thus briefly I have assembled whatever (apart from a few common plants) was seen on our rambles, so that it may be known to our friends and others interested in these matters how much labour and money we expended to advance the study of botany. And because (please take note !) our efforts were directed to utility rather than to ostentation, we replaced rare specimens after examining them in detail. But this year's work is merely a prelude to the strenuous years to come, for the success of which we invoke the blessing of God. Amen.

A further list of the plants of the Hampstead Heath area is to be found in Johnson (1632), the two together forming the first relatively complete list of the plants of a restricted area to be published in Britain. In the list which follows some twenty plants which also occurred in the 1629 list have been omitted.

Red Goosefoot (*Chenopodium rubrum* L.).
Stinking Goosefoot (*C. vulvaria* L.).
†Ivy-leaved Speedwell (*Veronica hederifolia* L.).
Thyme-leaved Sandwort (*Arenaria serpyllifolia* L.).
†Bur-Marigold (*Bidens tripartitus* L.).
Wall Rocket (*Diplotaxis tenuifolia* (L.) DC.).
Burr Chervil (*Anthriscus vulgaris* Bernh.).
Common Whitlow Grass (*Erophila verna* (L.) E. Meyer).
Rue-leaved Saxifrage (*Saxifraga tridactylites* L.).
Small Fleabane (*Pulicaria vulgaris* Gaertn.).
†Great Bindweed (*Convolvulus sepium* L.).
†Prickly Lettuce (*Lactuca serriola* L.).
Wild Lettuce (*L. virosa* L.).
Spotted Medick (*Medicago arabica* (L.) All.).
†Wild Rose (*Rosa* sp.).
Burnet Rose (*R. spinosissima* L.).
Common Valerian (*Valeriana sambucifolia* Mikan).
Broad-leaved Pondweed (*Potamogeton natans* L.).
Lousewort (*Pedicularis sylvatica* L.).
Ivy-leaved Crowfoot (*Ranunculus hederaceus* L.).
†Darnel (*Lolium temulentum* L.).
Vernal Sedge (*Carex caryophylla* Latour).
Fox Sedge (*C. otrubæ* Podpera.)
Common Quaking Grass (*Briza media* L.).
†Field Woodrush (*Luzula campestris* (L.) Willd.).
Mat-grass (*Nardus stricta* L.).
Toad Rush (*Juncus bufonius* L.).
Jointed Rush (*J. acutiflorus* (Ehrh.) Hoffm.).
Bulbous Buttercup (*Ranunculus bulbosus* L.).
Fumitory (*Fumaria officinalis* L.).
†Ash (*Fraxinus excelsior* L.).
†Oak (*Quercus robur* L.).
†Bramble (*Rubus fruticosus* L.).
†Elder (*Sambucus nigra* L.).
Privet (*Ligustrum vulgare* L.).
†Hazel (*Corylus avellana* L.).

Ling (*Calluna vulgaris* (L.) Hull).
†Hawthorn (*Cratægus oxyacantha* L.).
Dyer's Greenweed (*Genista tinctoria* L.).
†Gorse or Furze (*Ulex europæus* L.).
Juniper (*Juniperus communis* L.).
[Osier (*Salix viminalis* L.)].
†[Sallow (*S. caprea* L.)].
Bullace (*Prunus institia*) L.).
†Blackthorn (*P. spinosa* L.).
Early Purple Orchis ('*Orchis mascula* L.).
Spotted Orchis (*O. maculata* L.).
†Earth-nut or Pig-nut (*Conopodium majus* (Gouan) Loret & Barrandon).
Wild Chamomile (*Matricaria chamomilla* L.).
†Wood Club-rush (*Scirpus sylvaticus* L.).
Crosswort (*Galium cruciata* (L.) Scop.).
Musk Mallow (*Malva moschata* L.).
Grass Vetchling (*Lathyrus nissolia* L.).
Common Twayblade (*Listera ovata* (L.) R.Br.).
Royal Fern (*Osmunda regalis* L.).
Bogbean (*Menyanthes trifoliata* L.).
Great Broomrape (*Orobanche rapum-genistæ* Thuill.).
Sheep's-bit Scabious (*Jasione montana* L.).
Lesser Skullcap (*Scutellaria minor* Huds.).
Meadow Thistle (*Cirsium anglicum* (Lam.) DC.).

THE BIRDS OF THE LONDON AREA

(From Fitter & Parrinder (1944))

G :- General, found in most suitable habitats.

L. :- Local, irregularly distributed over suitable habitats, and usually found in less than half of them.

S. :- Scarce, not more than ten individuals or breeding pairs in the area at any one time.

O. :- Occasional, has been recorded more than five, but fewer than twenty times in 1924-43.

* Also winter visitor.   † Also passage migrant.   ‡ Has bred since 1920.

I.—*Breeding species resident throughout the year* (74).

Carrion-Crow (*Corvus corone corone* L.), G.
Rook (*Corvus f. frugilegus* L.), G.
Jackdaw (*Corvus monedula spermologus* Vieill.), G.
Magpie (*Pica p. pica* (L.)), G.
British Jay (*Garrulus glandarius rufitergum* Hart.), G.
Starling (*Sturnus v. vulgaris* L.), G.
Hawfinch (*Coccothraustes c. coccothraustes* (L.)), G.
Greenfinch (*Chloris ch. chloris* (L.)), G.
British Goldfinch (*Carduelis c. britannica* (Hart.)), G.
*Lesser Redpoll (*Carduelis flammea cabaret* (P. L. S. Müll.), L.
Linnet (*Carduelis cannabina cannabina* (L.)), G.
British Bullfinch (*Pyrrhula p. nesa* Math. & Ired.), G.
British Chaffinch (*Fringilla coelebs gengleri* Kleinschm.), G.
Corn-Bunting (*Emberiza calandra* L.), L.
Yellow Bunting (*Emberiza c. citrinella* L.), G.
Cirl-Bunting (*Emberiza c. cirlus* L.), S.
Reed-Bunting (*Emberiza s. schœniclus* (L.)), G.
House-Sparrow (*Passer d. domesticus* (L.)), G.
Tree-Sparrow (*Passer m. montanus* (L.)), L.
Wood-Lark (*Lullula a. arborea* (L.)), S.
Sky-Lark (*Alauda arvensis arvensis* L.), G.

*Meadow-Pipit (*Anthus pratensis* (L.)), L.
*Grey Wagtail (*Motacilla c. cinerea* Tunst.), S.
Pied Wagtail (*Motacilla alba yarrellii* Gould), G.
British Tree-Creeper (*Certhia familiaris britannica* Ridgw.), G.
British Nuthatch (*Sitta europæa affinis* Blyth), G.
British Great Tit (*Parus major newtoni* Prazak), G.
British Blue Tit (*Parus cæruleus obscurus* Prazak), G.
British Coal-Tit (*Parus ater britannicus* Sharpe & Dress.), G.
British Marsh-Tit (*Parus palustris dresseri* Stejn.), G.
British Willow-Tit (*Parus atricapillus kleinschmidti* Hellm.), L.
British Long-tailed Tit (*Aegithalos caudatus rosaceus* Mathews), G.
British Goldcrest (*Regulus r. anglorum* Hart.), G.
*Dartford Warbler (*Sylvia undata dartfordiensis* Lath.), S.
Mistle-Thrush (*Turdus v. viscivorus* L.), G.
British Song-Thrush (*Turdus e. ericetorum* Turton), G.
Blackbird (*Turdus m. merula* L.), G.
British Stonechat (*Saxicola torquata hibernans* (Hart.)), L.
British Robin (*Erithacus rubecula melophilus* Hart.), G.
British Hedge-Sparrow (*Prunella modularis occidentalis* (Hart.)), G.
Wren (*Troglodytes t. troglodytes* (L.)), G.
Kingfisher (*Alcedo atthis ispida* L.), G.
Green Woodpecker (*Picus viridis pluvius* Hart.), G.
British Great Spotted Woodpecker (*Dryobates major anglicus* (Hart.), G.
British Lesser Spotted Woodpecker (*Dryobates minor comminutus* (Hart.), L.
Little Owl (*Athene noctua vidalii* A. E. Brehm), G.
Long-eared Owl (*Asio o. otus* (L.)), S.
British Tawny Owl (*Strix aluco sylvatica* Shaw), G.
White-breasted Barn-Owl (*Tyto a. alba* (Scop.), L.
Kestrel (*Falco t. tinnunculus* L.), G.
Sparrow-Hawk (*Accipiter n. nisus* (L.)), G.
Common Heron (*Ardea c. cinerea* L.), L.
Mute Swan (*Cygnus olor* (Gm.)), G.
Canada Goose (*Branta c. canadensis* (L.)), L.
Mallard (*Anas p. platyrhyncha* L.), G.
Gadwall (*Anas strepera* L.), S.
*Teal (*Anas. c. crecca* L.), S.
*Common Pochard (*Aythya ferina* (L.)), S.
*Tufted Duck (*Aythya fuligula* (L.)), L.
Great Crested Grebe (*Podiceps c. cristatus* (L.)), G.
Little Grebe or Dabchick (*Podiceps r. ruficollis* (Pall.)), G.
Wood-Pigeon (*Columba p. palumbus* L.), G.
Stock-Dove (*Columba œnas* L.), G.
London Pigeon (*Columba*), G.

*Woodcock (*Scolopax rusticola* L.), L.
Common Snipe (*Capella g. gallinago* (L.)), G.
British Redshank (*Tringa totanus britannica* Math.), G.
Lapwing (*Vanellus vanellus* (L.),), G.
*Water-Rail (*Rallus a. aquaticus* L.), S.
Moorhen (*Gallinula ch. chloropus* (L.),), G.
Coot (*Fulica a. atra* L.), G.
Pheasant (*Phasianus colchicus* L.), G.
Common Partridge (*Perdix p. perdix* (L.)), G.
Red-legged Partridge (*Alectoris r. rufa* (L.)), L.

II.—*Breeding species resident in summer only* (26).

Tree-Pipit (*Anthus t. trivialis* (L.)), G.
Yellow Wagtail (*Motacilla flava flavissima* (Blyth)), L.
Red-backed Shrike (*Lanius c. collurio* L.), L.
Spotted Flycatcher (*Muscicapa s. striata* (Pall.)), G.
Chiffchaff (*Phylloscopus c. collybita* (Vieill.)), G.
Willow-Warbler (*Phylloscopus t. trochilus* (L.)), G.
Wood-Warbler (*Phylloscopus sibilatrix* (Bechst.)), G.
Grasshopper-Warbler (*Locustella n. nævia* (Bodd.)), L.
Reed-Warbler (*Acrocephalus s. scirpaceus* (Herm.)), L.
Sedge-Warbler (*Acrocephalus schœnobænus* (L.)), G.
Garden-Warbler (*Sylvia borin* (Bodd.)), G.
Blackcap (*Sylvia a. atricapilla* (L.)), G.
Whitethroat (*Sylvia c. communis* Lath.), G.
Lesser Whitethroat (*Sylvia c. curruca* (L.)), G.
Whinchat (*Saxicola rubetra* (L.)), L.
Redstart (*Phœnicurus ph. phœnicurus* (L.)), L.
†Black Redstart (*Phœnicurus ochrurus gibraltariensis* (Gm.),) S.
Nightingale (*Luscinia m. megarhyncha* Brehm), L.
Swallow (*Hirundo r. rustica* L.), G.
House-Martin (*Delichon u. urbica* (L.)), G.
Sand-Martin (*Riparia r. riparia* (L.)), G.
Swift (*Apus a. apus* (L.)), G.
Nightjar (*Caprimulgus e. europæus* L.), G.
Wryneck (*Jynx t. torquilla* L.), L.
Cuckoo (*Cuculus canorus canorus* L.), G.
Turtle-Dove (*Streptopelia t. turtur* (L.)), G.

III.—*Winter visitors* (45).

Hooded Crow (*Corvus cornix cornix* L.), S.
Siskin (*Carduelis spinus* (L.)), L.

†Mealy Redpoll (*Carduelis f. flammea* (L.)), O.

‡Common Crossbill (*Loxia c. curvirostra* L.), S.

†Continental Chaffinch (*Fringilla c. cœlebs* L.), O.

Brambling (*Fringilla montifringilla* L.), L.

Rock-Pipit (*Anthus spinoletta petrosus* (Mont.)), S.

Great Grey Shrike (*Lanius e. excubitor* L.), O.

Waxwing (*Bombycilla g. garrulus* (L.)), O.

Continental Goldcrest (*Regulus r. regulus* (L.)), O.

Firecrest (*Regulus i. ignicapillus* (Temm.), O.

Fieldfare (*Turdus pilaris* L.), G.

†Continental Song-Thrush (*Turdus ericetorum philomelus* Brehm), O.

Redwing (*Turdus m. musicus* L.), G.

†Continental Hedge-Sparrow (*Prunella m. modularis* (L.)), O.

Short-eared Owl (*Asio f. flammeus* (Pontopp.)), O.

†Sheld-Duck (*Tadorna tadorna* (L.)), S.

Wigeon (*Anas penelope* L.), G.

Pintail (*Anas acuta acuta* L.), S.

Scaup-Duck (*Aythya m. marila* (L.)), S.

Goldeneye (*Bucephala c. clangula* (L.), G.

Long-tailed Duck (*Clangula hyemalis* (L.)), O.

Goosander (*Mergus merganser merganser* L.), G.

Red-breasted Merganser (*Mergus serrator* L.), S.

Smew (*Mergus albellus* L.), G.

Cormorant (*Phalacrocorax c. carbo* (L.)) , L.

†Shag (*Phalacrocorax a. aristotelis* (L.)), S.

Red-necked Grebe (*Podiceps g. griseigena* (Bodd.)), S.

Slavonian Grebe (*Podiceps auritus* (L.)), S.

Great Northern Diver or Loon (*Colymbus immer* Brünn.), O.

Black-throated Diver (*Colymbus a. arcticus* L.), O.

Red-throated Diver (*Colymbus stellatus* Pontopp.), S.

Jack Snipe (*Lymnocryptes minimus* (Brünn.)), L.

†Dunlin (*Calidris alpina* (L), ? subsp.), L.

†Green Sandpiper (*Tringa ochropus* L.), S.

Golden Plover (*Pluvialis apricaria* (L.), ? su sp.), L.

Black-headed Gull (*Larus r. ridibundus* L.), G.

Common Gull (*Larus c. canus* L.), G.

Herring-Gull (*Larus a. argentatus* Pont.), G.

Scandinavian Lesser Black-backed Gull (*Larus f. fuscus* L.), **L.**

Great Black-backed Gull (*Larus marinus* L.), L.

Glaucous Gull (*Larus hyperboreus* Gunn.), O.

Iceland Gull (*Larus glaucoides* Meyer), O.

†Kittiwake (*Rissa t. tridactyla* (L.)), S.

Little Auk (*Alle a. alle* (L.)), O.

IV.—*Passage Migrants* (41).

Snow Bunting (*Plectrophenax n. nivalis* (L.)), O.
‡Blue-headed Wagtail (*Motacilla f. flava* L.), O.
White Wagtail (*Motacilla a. alba* L.), S.
Pied Flycatcher (*Muscicapa h. hypoleuca* (Pall.)), S.
Wheatear (*Oenanthe œ. œnanthe* (L.)), G.
Peregrine Falcon (*Falco p. peregrinus* Tunst.), S.
Hobby (*Falco s. subbuteo* L.), O.
*Merlin (*Falco columbarius æsalon* Tunst.), O.
Common Buzzard (*Buteo b. buteo* (L.)), O.
Garganey (*Anas querquedula* L.), O.
*Shoveler (*Spatula clypeata* (L.)), L.
*Common Scoter (*Melanitta n. nigra* (L.)), S.
Velvet Scoter (*Melanitta f. fusca* (L.)), O.
*Black-necked Grebe (*Podiceps n. nigricollis* Brehm), L.
Black-tailed Godwit (*Limosa l. limosa* (L.)), O.
*Common Curlew (*Numenius a. arquata* (L.)), G.
Whimbrel (*Numenius ph. phæopus* (L.)), O.
Grey Phalarope (*Phalaropus fulicarius* (L.)), O.
Turnstone (*Arenaria i. interpres* (L.)), O.
Knot (*Calidris canutus canutus* (L.)), O.
Curlew-Sandpiper (*Calidris testacea* (Pall.)), O.
Little Stint (*Calidris minuta* (Leisl.)), O.
Sanderling (*Crocethia alba* (Pall.)), O.
Ruff (*Philomachus pugnax* (L.)), S.
Common Sandpiper (*Actitis hypoleucos* (L.)), G.
Wood-Sandpiper (*Tringa glareola* L.), O.
Greenshank (*Tringa nebularia* Gunn.), L.
*Ringed Plover (*Charadrius hiaticula* L. ? subsp,) L.
*Grey Plover (*Squatarola squatarola* (L.)), O.
British Oyster-catcher (*Hæmatopus ostralegus occidentalis* Neum.), O.
Stone-Curlew (*Burhinus œ. œdicnemus* (L.)), O.
Black Tern (*Chlidonias n. niger* (L.)), L.
Sandwich Tern (*Sterna sandvicensis sandvicensis* Lath.), O.
Common Tern (*Sterna h. hirundo* L.), G.
Arctic Tern (*Sterna macrura* Naumann), O.
Little Tern (*Sterna a. albifrons* Pall.), O.
*Little Gull (*Larus minutus* Pall.), O.
British Lesser Black-backed Gull (*Larus fuscus grællsii* Brehm), G.
British Razorbill (*Alca torda britannica* Ticehurst), O.
Southern Puffin (*Fratercula arctica grabæ* (Brehm)), O.
Corncrake (*Crex crex* (L.)), S.

THE LOST PLANTS OF MIDDLESEX

(From Salisbury (1927))

———————

LIST OF PLANTS which have become extinct or seriously diminished (D) in Middlesex within recent years (55 species, of which 52 extinct).

Columbine (*Aquilegia vulgaris* L.).
London Rocket (*Sisymbrium irio* L.).
Proliferous Pink (*Dianthus prolifer* L.).
Deptford Pink (*D. armeria* L.).
Marsh St. John's Wort (*Hypericum elodes* L.).
Marsh Mallow (*Althæa officinalis* L.).
Fenugreek (*Trigonella purpurascens* Lam.).
Sulphur Clover (*Trifolium ochroleucum* L.).
Rough Clover (*T. scabrum* L.).
Clustered Clover (*T. glomeratum* L.).
Yellow Vetchling (*Lathyrus aphaca* L.).
Medlar (*Mespilus germanica* L.).
Hyssop Loosestrife (*Lythrum hyssopifolia* L.).
Common Sundew (*Drosera rotundifolia* L.) D.
Hare's-ear (*Bupleurum rotundifolium* L.).
Hartwort (*Tordylium maximum* L.).
Wall Bedstraw (*Galium anglicum* Huds.).
Star-thistle (*Centaurea calcitrapa* L.).
Willow Lettuce (*Lactuca saligna* L.).
Marsh Sow-thistle (*Sonchus palustris* L.).
Lamb's Succory (*Arnoseris pusilla* Gaertn.).
Rampion (*Campanula rapunculus* L.).
Bog Pimpernel (*Anagallis tenella* L.) D.
Common Bladderwort (*Utricularia vulgaris* L.).
Lesser Bladderwort (*U. minor* L.).
Marsh Gentian (*Gentiana pneumonanthe* L.).
Greater Dodder (*Cuscuta europæa* L.).
Green Hound's-tongue (*Cynoglossum montanum* L.).

Toothwort (*Lathræa squamaria* L.).
Water Figwort (*Scrophularia ehrharti* Stevens).
Water Mint (*Mentha pubescens* Willd.).
Whorled Mint (*M. gentilis* L.).
Pennyroyal (*M. pulegium* L.).
Rupture-wort (*Herniaria glabra* L. ; *H. hirsuta* L.).
Sea Orache (*Atriplex littoralis* L.).
Golden Dock (*Rumex maritimus* L.).
Bog-Myrtle (*Myrica gale* L.).
Bay Willow (*Salix pentandra* L.).
Juniper (*Juniperus communis* L.).
Star-fruit (*Damasonium alisma* Hill) D.
Military Orchis (*Orchis militaris* L. ; *O. purpurea* Huds.).
Dwarf Orchis (*O. ustulata* L.).
Pyramidal Orchis (*O. pyramidalis* L.).
Fragrant Orchis (*Habenaria conopsea* L.).
Bee Orchis (*Ophrys apifera* Huds.).
Summer Snowflake (*Leucoium æstivum* L.).
May-lily (*Maianthemum bifolium* Schmidt) (re-introduced).
Brown Sedge (*Cyperus fuscus* L.).
Tufted Sedge (*Scirpus cæspitosus* L.).
Bulrush (*S. tabernæmontani* Gm.).
Perennial Oat (*Avena pubescens* Huds.).
Pillwort (*Pilularia globulifera* L.).
Royal Fern (*Osmunda regalis* L.).
Rusty-back Fern (*Ceterach officinarum* DC.).

THE HUNDRED COMMONEST LONDON PLANTS

PLANTS which have been recorded in all or almost all the twenty-four districts into which the London Natural History Society divides the London area for recording purposes ; they are thus the most widely distributed rather than most numerically abundant plants.

(From Bishop *et al.* (1928-36))

Meadow Buttercup (*Ranunculus acer* L.).
Creeping Buttercup (*R. repens* L.).
Lesser Celandine (*R. ficaria* L.).
Hedge Mustard (*Sisymbrium officinale* (L.) Scop.).
Jack-by-the-Hedge (*Alliaria officinalis* Andrz.).
Shepherd's Purse (*Capsella bursa-pastoris* (L.) Medik).
White Campion (*Lychnis alba* Mill.).
Mouse-ear Chickweed (*Cerastium vulgatum* L.).
Chickweed (*Stellaria media* Vill.).
Lesser Stitchwort (*S. graminea* L.).
Dove's-foot Cranesbill (*Geranium molle* L.).
Cut-leaved Cranesbill (*G. dissectum* L.).
Black Medick (*Medicago lupulina* L.).
Red Clover (*Trifolium pratense* L.).
Dutch Clover (*T. repens* L.).
Lesser Yellow Trefoil (*T. dubium* Sibth.).
Bird's-foot Trefoil (*Lotus corniculatus* L.).
Common Vetch (*Vicia angustifolia* L.).
Meadow Vetchling (*Lathyrus pratensis* L.).
Bramble (*Rubus fruticosus* L.).
Avens (*Geum urbanum* L.).
Cinquefoil (*Potentilla reptans* L.).
Silverweed (*P. anserina* L.).
Parsley Piert (*Alchemilla arvensis* (L) Scop.).
Common Agrimony (*Agrimonia eupatoria* L.).

Dog·Rose (*Rosa canina* L. agg.).
Hawthorn (*Cratægus oxyacantha* L.).
Rose-bay (*Epilobium angustifolium* L.).
White Bryony (*Bryonia dioica* Jacq.).
Goutweed (*Aegopodium podagraria* L.).
Wild Chervil (*Anthriscus sylvestris* Hoffm.).
Fool's Parsley (*Aethusa cynapium* L.).
Hogweed (*Heracleum sphondylium* L.).
Wild Carrot (*Daucus carota* L.).
Ivy (*Hedera helix* L.).
Elder (*Sambucus nigra* L.).
Honeysuckle (*Lonicera periclymenum* L.).
Ladies' Bedstraw (*Galium verum* L.).
Goosegrass (*G. aparine* L.).
Common Teasel (*Dipsacus sylvestris* Huds.).
Field Scabious (*Scabiosa arvensis* L.).
Daisy (*Bellis perennis* L.).
Yarrow (*Achillea millefolium* L.).
Ox-eye Daisy (*Chrysanthemum leucanthemum* L.).
Scentless Mayweed (*Matricaria inodora* L.).
Coltsfoot (*Tussilago farfara* L.).
Groundsel (*Senecio vulgaris* L.).
Ragwort (*S. jacobæa* L.).
Spear Thistle (*Cirsium lanceolatum* (L) Scop.).
Marsh Thistle (*C. palustre* (L.) Scop.).
Field Thistle (*C. arvense* (L.) Scop.).
Black Knapweed (*Centaurea nigra* L.).
Nipplewort (*Lapsana communis* L.).
Mouse-ear Hawkweed (*Hieracium pilosella* L.).
Cat's-ear (*Hypochæris radicata* L.).
Autumnal Hawkbit (*Leontodon autumnalis* L.).
Dandelion (*Taraxacum officinale* Weber).
Sow-thistle (*Sonchus oleraceus* L.).
Scarlet Pimpernel (*Anagallis arvensis* L.)
Ash (*Fraxinus excelsior* L.).
Great Bindweed (*Convolvulus sepium* L.).
Field Bindweed (*C. arvensis* L.).
Bittersweet (*Solanum dulcamara* L.).
Yellow Toadflax (*Linaria vulgaris* Mill.).
Field Speedwell (*Veronica polita* Fr.).
Buxbaum's Speedwell (*Veronica persica* Poir.).
Wall Speedwell (*V. arvensis* L.).
Germander Speedwell (*V. chamædrys* L.).

Ground Ivy (*Nepeta hederacea* (L.) Trev.).
Self-heal (*Prunella vulgaris* L.).
Hedge Woundwort (*Stachys sylvatica* L.).
Red Dead-Nettle (*Lamium purpureum* L.).
White Dead-Nettle (*L. album* L.).
Black Horehound (*Ballota nigra* L.).
Ribwort Plantain (*Plantago lanceolata* L.).
Greater Plantain (*P. major* L.).
Black Bindweed (*Polygonum convolvulus* L.).
Knotgrass (*P. aviculare* L.).
Persicaria (*P. persicaria* L.).
Common Dock (*Rumex obtusifolius* L.).
Curled Dock (*R. crispus* L.).
Sorrel (*R. acetosa* L.).
Sheep's Sorrel (*R. acetosella* L.).
Petty Spurge (*Euphorbia peplus* L.).
Elm (*Ulmus procera* Sabob.).
Stinging Nettle (*Urtica dioica* L.).
Sallow (*Salix caprea* L.).
Lords and Ladies (*Arum maculatum* L.).
Fox Sedge (*Carex otrubæ* Podpera).
Marsh Fox-tail Grass (*Alopecurus geniculatus* L.).
Meadow Fox-tail Grass (*A. pratensis* L.).
Yorkshire Fog (*Holcus lanatus* L.).
False Oat (*Arrhenatherum elatius* (L.) Beauv.).
Cock's-foot Grass (*Dactylis glomerata* L.).
Annual Meadow Grass (*Poa annua* L.).
Meadow Grass (*P. pratensis* L.).
Rye-grass (*Lolium perenne* L.).
Couch-grass (*Agropyron repens* (L.) Beauv.).
Wall Barley (*Hordeum murinum* L.).
Bracken (*Pteridium aquilinum* (L.) Kuhn).

NATURAL HISTORY SOCIETIES, MUSEUMS AND LOCALITIES IN THE LONDON AREA

COUNTY NATURAL HISTORY SOCIETIES

LONDON NATURAL HISTORY SOCIETY (founded 1858). (Hon Secretary, H. A. Toombs, British Museum (Natural History), S.W.7.) The only natural history society covering the whole London area within twenty miles of St. Paul's Cathedral. Indoor meetings are held at the London School of Hygiene and Tropical Medicine, Keppel St., W.C.1., and outdoor meetings in all parts of the country round London. The Society's journals are *The London Naturalist* and *The London Bird Report*, published annually. There are eight sections (Archæology, Botany, Ecology, Entomology, Geology, Ornithology, Plant Galls, and Ramblers), and a branch at Chingford, Essex. Natural history records for the London area may be sent to the Society's recorders as under :

Mammals, Birds, Reptiles, Amphibians—R. S. R. Fitter, 39 South Grove House, Highgate, N.6.

Botany—J. E. Lousley, 7 Penistone Road, S.W.16.

Plant Galls—H. J. Burkill, 3 Newman's Court, E.C.3.

SOUTH LONDON ENTOMOLOGICAL AND NATURAL HISTORY SOCIETY (founded 1872). (Hon. Secretary, F. Stanley-Smith, Hatch House, Pilgrim's Hatch, Brentwood, Essex.) Mainly an entomological society ; indoor meetings are held at the Chapter House of Southwark Cathedral. The Society publishes *Proceedings* and *Transactions*.

ESSEX FIELD CLUB (founded 1880). (Hon. Secretary, Percy G. Thompson, Essex Museum of Natural History, Romford Road, Stratford, E.15.) The county natural history society for Essex ; maintains museums at Stratford and at Queen Elizabeth's Lodge, Chingford ; publishes *The Essex Naturalist*.

HERTFORDSHIRE NATURAL HISTORY SOCIETY (founded 1875). (Hon. Secretary, Miss Eileen Gibbs, Houndspath, St. Albans, Herts.) The county natural history society for Hertfordshire ; meetings are usually held at St. Albans. The Society publishes *Transactions*. The principal recorders are as follows :

Birds—H. H. S. Hayward, 60 Ridge Crest, Enfield, Middlesex.
Butterflies and Moths—Dr. A. H. Foster, 13 Tilehouse Street, Hitchin, Herts.
Botany—Dr. E. J. Salisbury, Royal Botanic Gardens, Kew, Surrey.

BRITISH EMPIRE NATURALISTS ASSOCIATION (founded 1905). (Hon. Organising Secretary, Leslie Beckett, 22 South Drive, Ruislip, Middlesex.) Aims " to bring naturalists and nature-lovers in all parts of the Empire into helpful communication ; to secure the protection of wild life and objects of interest to the nature student ; and to promote the preservation of the natural beauties of the countryside." Publishes *Countryside* quarterly ; has a London branch.

SOUTH-EASTERN UNION OF SCIENTIFIC SOCIETIES (founded 1896). (Hon. Secretaries, A. Farquharson, Le Play House, Albert Road, Malvern, Worcs. ; F. J. Epps, 78 Dunwich Road, Bexleyheath, Kent.) A union of the natural history and scientific societies of South-eastern England. Publishes *The South-Eastern Naturalist and Antiquary* ; holds annual congresses. There are five sections : Archæology, Botany, Geology, Social Science, Zoology.

LOCAL NATURAL HISTORY SOCIETIES

BARNET AND DISTRICT NATURAL HISTORY SOCIETY (founded 1905). (Hon. Secretary, R. J. Griffiths, 7 Netherlands Road, New Barnet, Herts.)

CROYDON NATURAL HISTORY AND SCIENTIFIC SOCIETY (founded 1870). (Hon. Secretary, C. T. Prime, 2 Lansdown Road, Croydon, Surrey.)

GRAVESEND SOCIETY FOR ARCHÆOLOGY, SCIENCE, LITERATURE AND ART (founded 1926). (Hon. Secretary, A. J. Philip, Public Library, Gravesend, Kent.)

HAMPSTEAD SCIENTIFIC SOCIETY (founded 1899). (Hon. Secretary, Mrs. H. Baily, 74 Lawn Road, Hampstead, N.W.3.)

HOLMESDALE NATURAL HISTORY CLUB (founded 1857). (Hon. Secretary, Miss D. Powell, Aldersyde, Reigate, Surrey.)

PLUMSTEAD AND DISTRICT NATURAL HISTORY SOCIETY (founded 1927). (Hon. Secretary, c/o Librarian, Museum and Library, 232 Plumstead High Street, S.E.18.)

PURLEY NATURAL HISTORY AND SCIENTIFIC SOCIETY (founded 1923). (Hon. Secretary, H. F. Haskins, 39 Windermere Road, Coulsdon, Surrey.)

SIDCUP LITERARY AND SCIENTIFIC SOCIETY (founded 1880.) (Hon. Secretary, C. S. Bryant, 4 Priestlands Road, Sidcup, Kent.)

SOUTH LONDON BOTANICAL INSTITUTE (founded 1911). (Hon. Secretary, W. R. Sherwin, 323 Norwood Road, Herne Hill, S.E.24.)

STREATHAM ANTIQUARIAN AND NATURAL HISTORY SOCIETY (founded 1933).

(Hon. Secretary, Col. Sir G. R. Hearn, 52 Woodbourne Avenue, Streatham, S.W.16.)

WEST KENT SCIENTIFIC SOCIETY (founded 1859). (Hon. Secretary, C. C. Newell, 52 Hardy Road, Blackheath, S.E.3.)

WIMBLEDON NATURAL HISTORY SOCIETY OF THE JOHN EVELYN CLUB (founded 1932). (Hon. Secretary, C. P. Castell, 52 Graham Road, Wimbledon, S.W.19.)

WOOLWICH HISTORICAL AND SCIENTIFIC SOCIETY (founded 1921). (Hon Secretary, C. Foster, 71 Rectory Place, Woolwich, S.E.18.)

MUSEUMS WITH NATURAL HISTORY COLLECTIONS

(from the Directory of Museums and Art Galleries in the British Isles. Compiled by the Museums Association, 1931.)

BRITISH MUSEUM (NATURAL HISTORY), Cromwell Road, South Kensington, S.W.7. The national collection and natural history museum is open free ; apart from one case of London birds, it has no special London collections.

County of London

BETHNAL GREEN. Bethnal Green Museum, Cambridge Road, E.2 has section on local topography and collection of British and foreign birds to aid nature study in East London.

FOREST HILL. Horniman Museum, London Road, S.E.23 has a large range of general natural history exhibits.

HAMPSTEAD. Hampstead Museum, Arkwright Road, N.W.3 contains local archæological and natural history specimens.

ST. GEORGE'S IN THE EAST. St. George's Nature Study Museum, Cable Street, E.1 has living specimens of vertebrates, molluscs and insects, also a beehive, aquaria and exhibits.

WHITECHAPEL. Whitechapel Museum, Whitechapel High Street, E.1 has zoological collections.

WOOLWICH. Woolwich Borough Museum contains local shells, British and foreign birds, eggs of lepidoptera, zoological specimens.

Essex

CHINGFORD. Epping Forest Museum, Queen Elizabeth's Lodge : natural history and archæology of Epping Forest and the Lea and Roding valleys.

STRATFORD. Essex Museum of Natural History, Romford Road, E15 : Essex natural history exhibits.

Hertfordshire

HERTFORD. Hertford Museum, 18 Bull Plain : some local geological and zoological specimens.

ST. ALBANS. Hertfordshire County Museum: zoological (birds, lepidoptera, beetles), botanical (including mosses and lichens) and geological collections.

Kent

BEXLEY HEATH. Bexley Heath Museum, Danson Park : general and local collections of shells, birds' eggs and lepidoptera.

Middlesex

SOUTHGATE. Broomfield House : small collection of local flora and fauna.

TWICKENHAM. Twickenham Museum, York House : collection illustrating fauna of Thames gravels.

Surrey

KEW. Museums of Economic Botany, Royal Botanic Gardens : collections on a world scale.

KINGSTON-ON-THAMES. Kingston-on-Thames Museum : Thames valley birds.

REIGATE. Holmesdale Natural History Club Museum : collections of British birds' eggs, insects, plants, fossils, etc.

WIMBLEDON. Museum of the John Evelyn Club for Wimbledon : local natural history collections.

There is, unfortunately, no museum in London with any comprehensive collections illustrating the natural history of the London area.

SOME NATURAL HISTORY LOCALITIES IN THE LONDON AREA

(Localities for which permits are needed are omitted.)

Essex

EPPING FOREST.—Badgers, red squirrels, fallow deer ; hawfinches, redstarts, nightingales, nightjars ; hornbeams.

RODING VALLEY (Ongar to Woodford). Snipe, redshanks, green sandpipers.

WALTHAMSTOW RESERVOIRS (accessible from Coppermill Lane and North Circular Road). Otters ; carrion-crow roost, heronry, waterfowl.

Hertfordshire

BRICKET WOOD COMMON. Warblers and nightingales ; one of the best southern examples of varied scrub successional to woodland.

CASSIOBURY PARK AND WHIPPENDELL WOODS (Watford). Woodland birds, aquatic birds on R. Gade.

COLNE VALLEY GRAVEL PITS (Hamper Hill to West Hyde). Waterfowl ; aquatic plants.

CUFFLEY GREAT WOODS. Woodland birds, esp. redstarts and nightjars.

ELSTREE (OR ALDENHAM) RESERVOIR. Waterfowl ; aquatic plants.

Kent

HIGH ELMS PARK AND CUCKOO WOOD (Downe). Typical beechwood ; includes Darwin's orchid bank ; Darwin's home, Down House, is in Downe village nearby, and is preserved by the British Association ; it is open to the public.

HAYES AND KESTON COMMONS. Good examples of oak-birch heath ; a fine bog flora on Keston Common.

LULLINGSTONE PARK. Deer ; orchids.

NORTH DOWNS (esp. the scarp and Cudham and Knockholt areas). Typical chalk butterflies and plants ; orchids.

Middlesex

BRENT RESERVOIR (Hendon). Waterfowl.

BUSHY PARK. Deer ; redstarts ; horse-chestnuts.

HAREFIELD MOOR. Marsh plants.

HARROW WEALD AND STANMORE COMMONS. Warblers ; oak-birch woods.

KEN WOOD. Badgers ; may-lily (*Maianthemum bifolium*), fine beeches.

RUISLIP RESERVOIR (and adjacent Common, Copse and Park Woods). Willow-tits (in winter), grasshopper and other warblers, nightjars ; aquatic and woodland plants.

SCRATCH WOOD (Elstree). Woodland birds, esp. warblers.

STAINES RESERVOIRS (public causeway). Gull roost, waterfowl, esp. wigeon, goldeneye, goosander.

Surrey

ASHTEAD AND EPSOM COMMONS. Typical commons of the London Clay ; grasshopper and other warblers, nightingales, nightjars ; rich in lepidoptera.

BOOKHAM COMMON. Similar to Ashtead and Epsom Commons.

ESHER COMMON AND OXSHOTT HEATH. Typical heathland and pinewood area ; Black Pond and surrounds are rich in insects ; good for fungi.

MOLE VALLEY (Burford Bridge to Fetcham). Good chalk downland, beech, yew, box.

NORTH DOWNS (esp. scarp, e.g., Box Hill, Colley Hill and Marden Park). Butterflies and plants of the chalk, including orchids.

RICHMOND PARK AND WIMBLEDON COMMON. Badgers ; redstarts, heronry, waterfowl on Pen Ponds ; bog flora on Wimbledon Common.

SELSDON WOOD. Bird sanctuary, willow-tits and warblers.

THAMES TOWPATH (Richmond to Putney) and Lonsdale Road Reservoirs. Waterfowl, esp. gadwall and smew.

WALTON HEATH. Heathland birds and plants.

N.B.—Neither the list of localities nor the animals and plants to be found in them are intended to be exhaustive ; they merely illustrate the more interesting places for naturalists to visit round London, and the type of wild life to be found in them. Great care should, of course, be taken not to disturb any of these animals or plants in their few remaining wild localities near London.

LIST OF FLOWERING PLANTS AND FERNS RECORDED FROM BOMBED SITES IN LONDON

By E. J. Salisbury, D.Sc., F.R.S., Director of the Royal
Botanic Gardens, Kew

ONLY species that were fresh introductions are recorded. Bombed sites which included gardens or parts of gardens were not utilised unless the species present prior to bombing were known and could be excluded from the enumeration. In all 126 species are listed, all from sites within the County of London. The percentages in brackets indicate the proportion of sites on which the commoner species were found.

Bulbous Buttercup (*Ranunculus bulbosus* L.).
Creeping Buttercup (*R. repens* L.).
Field Poppy (*Papaver rhœas* L.).
Opium Poppy (*P. somniferum* L.).
Yellow Corydalis (*Corydalis lutea* (L.) DC.).
Marsh Watercress (*Rorippa islandica* (Oeder) Schinz Thellung.)
Hedge Mustard (*Sisymbrium officinale* (L.) Scop.).
Charlock (*Brassica arvensis* L.).
Penny Cress (*Thlaspi arvense* L.).
Shepherd's Purse (*Capsella bursa-pastoris* (L.) Medik. agg.), (15%).
Narrow-leaved Pepperwort (*Lepidium ruderale* L.).
Swine's cress (*Senebiera coronopus* Poir.).
Lesser Wart-cress (*S. didyma* Pers.).
Weld or Dyer's Rocket (*Reseda luteola* L.), (4%).
Bladder Campion (*Silene cucubalus* Wibel).
White Campion (*Lychnis alba* Mill.).
Red Campion (*L. dioica* L.).
Procumbent Pearlwort (*Sagina procumbens* L.).
Mouse-ear Chickweed (*Cerastium vulgatum* L.), (7%.).
Common Chickweed (*Stellaria media* Vill.), (26%).
Dwarf Mallow (*Malva rotundifolia* L.), (4%).

Common Mallow (*M. sylvestris* L.).
Musk Mallow (*M. moschata* L.).
Dove's-foot Cranesbill (*Geranium molle* L.).
Sycamore (*Acer pseudo-platanus* L.).
Gorse (*Ulex europæus* L.).
Broom (*Sarothamnus scoparius* (L.) Wimmer ex Koch.).
Black Medick (*Medicago lupulina* L.), (6%).
Common Melilot (*Melilotus officinalis* Lam.).
White Melilot (*M. albus* Medik.).
Red Clover (*Trifolium pratense* L.), (22%).
Dutch Clover (*T. repens* L.), (8%).
Alsike Clover (*T. hybridum* L.).
Hop Trefoil (*T. procumbens* L.).
Bird's-foot Trefoil (*Lotus corniculatus* L.).
Tufted Vetch (*Vicia cracca* L.).
Common Vetch (*V. sativa* L.).
Bramble (*Rubus fruticosus* L.).
Rose-bay Willow-herb (*Epilobium angustifolium* L.), (88%).
Codlins-and-cream (*E. hirsutum* L.).
Hoary Willow-herb (*E. parviflorum* Schreb.).
Broad-leaved Willow-herb (*E. montanum* L.), (6%).
Evening Primrose (*Œnothera biennis* L.).
Fool's Parsley (*Æthusa cynapium* L.).
Elder (*Sambucus nigra* L.), (5%).
Canadian Fleabane (*Erigeron canadensis* L.), (40%).
Feverfew (*Chrysanthemum parthenium* Bernh.).
Scentless Mayweed (*Matricaria inodora* L.), (16%).
Wild Chamomile (*M. chamomilla* L.).
Rayless Chamomile or Pineapple Weed (*M. discoidea* DC.).
Yarrow (*Achillea millefolium* L.), (4%).
Mugwort (*Artemisia vulgaris* L.), (8%).
Coltsfoot (*Tussilago farfara* L.), (65%).
Groundsel (*Senecio vulgaris* L.), (88%).
Sticky Groundsel (*S. viscosus* L.), (44%).
Oxford Ragwort (*S. squalidus* L.), (56%).
Ragwort (*S. jacobæa* L.).
Spear Thistle (*Cirsium lanceolatum* (L.) Scop.), (17%).
Creeping Thistle (*C. arvense* (L.) Scop.), (7%).
Cat's-ear (*Hypochæris radicata* L.), (4%).
Prickly Lettuce (*Lactuca serriola* L.).
Wild Lettuce (*L. virosa* L.), (6%).
Corn Sow-thistle (*Sonchus arvensis* L.).
Prickly Sow-thistle (*S. asper* (L.) Hill), (6%).

Common Sow-thistle (*S. oleraceus* L.), (44%).
Dandelion (*Taraxacum officinale* Weber), (28%).
Beaked Hawk's-beard (*Crepis taraxacifolia* Thuill.), (4%).
Smooth Hawk's-beard (*C. capillaris* Wallr.), (4%).
Nipplewort (*Lapsana communis* L.), (4%).
Gallant Soldiers (*Galinsoga parviflora* Cav.), (14%).
Scarlet Pimpernel (*Anagallis arvensis* L.).
Buddleia (*Buddleia davidii* Franchet=*B. variabilis* Hemsl.).
Lesser Bindweed (*Convolvulus arvensis* L.).
Great Bindweed (*C. sepium* L.).
Thorn-apple (*Datura stramonium* L.).
Bittersweet (*Solanum dulcamara* L.).
Black Nightshade (*S. nigrum* L.), (10%).
Tomato (*S. lycopersicum* L.), (12%).
Great Snapdragon (*Antirrhinum majus* L.).
Yellow Toadflax (*Linaria vulgaris* Mill.).
Ivy-leaved Toadflax (*L. cymbalaria* (L.), Mill.).
Musk Mimulus (*Mimulus moschatus* Dougl.).
Thyme-leaved Speedwell (*Veronica serpyllifolia* L.).
Field Speedwell (*V. polita* Fr.), (4%).
Black Horehound (*Ballota nigra* L.).
Greater Plantain (*Plantago major* L.), (16%).
Ribwort (*P. lanceolata* L.), (10%).
Many-seeded Goosefoot (*Chenopodium polyspermum* L.).
White Goosefoot or Fat-hen (*C. album* L.), (8%).
Red Goosefoot (*C. rubrum* L.), (6%).
Nettle-leaved Goosefoot (*C. murale* L.).
Common Orache (*Atriplex patula* L.), (6%).
Curled Dock (*Rumex crispus* L.), (10%).
Broad-leaved Dock (*R. obtusifolius* L.), (12%).
Marsh Dock (*R. limosus* Thuill.).
Sheep's Sorrel (*R. acetosella* L.).
Knotgrass (*Polygonum aviculare* L.), (16%).
Black Bindweed (*P. convolvulus* L.), (4%).
Persicaria (*P. persicaria* L.), (8%).
Garden Polygonum (*P. cuspidatum* Sieb. & Zucc.).
Petty Spurge (*Euphorbia peplus* L.).
Annual Mercury (*Mercurialis annua* L.).
Small Nettle (*Urtica urens* L.), (6%).
Stinging Nettle (*U. dioica* L.), (8%).
Sallow (*Salix caprea* L.), (16%).
Timothy-grass (*Phleum pratense* L.).
Meadow Foxtail-grass (*Alopecurus pratensis* L.).

Bent-grass (*Agrostis stolonifera* L.), (18%).
Cultivated Oat (*Avena sativa* L.).
Yorkshire Fog (*Holcus lanatus* L.), (28%).
Wall Barley (*Hordeum murinum* L.).
Barley (*Hordeum sativum* L.).
Couch-grass (*Agropyron repens* (L.) Beauv.).
Rye-grass (*Lolium perenne* L.), (46%).
Barren Brome-grass (*Bromus sterilis* L.).
Field Brome-grass (*B. hordeaceus* L.).
Sheep's Fescue (*Festuca rubra* L.).
Meadow Fescue (*F. pratensis* Huds.).
Cock's-foot grass (*Dactylis glomerata* L.), (10%).
Crested Dog's-tail (*Cynosurus cristatus* L.).
Annual Meadow Grass (*Poa annua* L.), (58%).
Flattened Meadow Grass (*P. compressa* L.).
Meadow Grass (*P. pratensis* L.), (4%).
Rough Meadow Grass (*P. trivialis* L.).
Male Fern (*Dryopteris filix-mas* (L.) Schott.).
Bracken (*Pteridium aquilinum* (L.) Khun.).

ABBOTT, W. J. L. (1892), The Section Exposed in the Foundations of the New Admiralty Offices. *Proc. Geol. Ass.*, *12*, 346-56.

ARDAGH, J. (1928), Ballard's " Catalogue of Islington Plants." *J. Bot.*, *66*, 185-94.

AUSTEN, MAJOR E. E. and HUGHES, A. W. McK. (1932), Clothes Moths and House-Moths : their life-history, habits and control. British Museum (Natural History), Economic Series No. 14.

BAKER, F. J. (1908), Mammals in *Victoria County History of Kent*, Vol. I.

BARRETT-HAMILTON G. E. H. and HINTON, M. A. C. (1910-21) A. History of British Mammals.

BARTLETT, T. L. (1944), Recoveries of Black-headed Gulls in Inner London. *London Bird Report*, 1943, 19.

BAYES, C. S. (1944), A Historical Sketch of Epping Forest. *London Naturalist*, 1943, 32-43.

BEADELL, A. (1932), Nature Notes of Warlingham and Chelsham. Croydon.

BEADNELL, SRGN. REAR-ADM. C. M. (1937), The Toll of Animal Life Exacted by Modern Civilisation. *Proc. Zool. Soc. Lond.*, A, *107*, 173-82.

BELL, W. G. (1924), The Plague.

BESANT, W. (1892), London.

BESANT, W. (1899), East London.

BISHOP, E. B., ROBBINS, R. W. and SPOONER, H. (1928-36), Botanical ˙ Records of the London Area. *London Naturalist*, 1927-35 (8 parts).

BLAKER, G. B. (1934), The Barn Owl in England and Wales. (Report of Inquiry by Royal Society for the Protection of Birds.)

BRETT-JAMES, N. G. (1935), The Growth of Stuart London.

BROGAN, D. W. (1943), The English People.

BUCKNILL, J. A. (1900), The Birds of Surrey.

BUCKNILL, J. A. and MURRAY, H. W. (1902), Mammalia in *Victoria County History of Surrey*, Vol. I.

BUXTON, E. N. (1901), Epping Forest.

CÆSAR, J. (1st cent. B.C.), De Bello Gallico.

CALVERT, G. W., FITTER, R. S. R. and HALE, R. W. (1944), Black Redstarts Breeding in Middlesex since 1926. *Brit. Birds*, *37*, 189-90.

CASSIUS DIO (1st-2nd cent. A.D.), Rhomaike Historia, LX.

CASTELL, C. P. (1945), The Bookham Common Survey, Third Year ; The Ponds and their Vegetation. *London Naturalist*, 1944.

CHRISTY, M. (1890), The Birds of Essex. Chelmsford.

COBBETT, W. (1821-32), Rural Rides.

COCKSEDGE, W. C. (1933), The Great North Wood. *London Naturalist*, 1932, 48-51.

COLLENETTE, C. L. (1937), A History of Richmond Park.

COLLENETTE, C. L. (1939), The Invertebrate Fauna of Hyde Park and Kensington Gardens : Insects. *London Naturalist*, 1938, 48-49.

COLLINGE, W. E. (1927), The Food of Some British Wild Birds.

COMMITTEE ON BIRD SANCTUARIES IN ROYAL PARKS (ENGLAND), (1929-39), Annual Reports, 1928-38.

CORNISH, C. J. (1902), The Naturalist on the Thames.

CROSSMAN, A. F. (1902), Mammalia in *Victoria County of Hertfordshire*, Vol. I.

CURTIS, W. (1777-98), Flora Londinensis : or Plates and Descriptions of such plants as grow wild in the Environs of London ; with their places of growth, and times of flowering ; their several names according to Linnæus and other authors : with a particular description of each plant in Latin and English ; to which are added their several uses in Medicine, Agriculture, Rural Economy and other Arts.

DARBY, H. C., ed. (1936), An Historical Geography of England before A.D. 1800. Cambridge.

DARWIN, C. (1859), The Origin of Species by Means of Natural Selection.

DAWSON, F. L. M. (1940), The Small Mammal Population of Mill Hill, with brief notes on the reptiles and amphibians. *London Naturalist*, 1939, 10-15.

DEFOE, DANIEL (1724), A Tour Thro' the whole Island of Great Britain.

DE QUINCEY, Thomas (1821), Confessions of an English Opium Eater.

DEWEY, H. (1932), The Palæolithic Deposits of the Lower Thames Valley. *Quart. J. Geol. Soc.*, 88, 35-56.

DOMESDAY BOOK FOR MIDDLESEX (1086).

DREWITT, F. Dawtrey (1928), The Romance of the Apothecaries' Garden at Chelsea. Cambridge.

DURRANT, J. H. and BEVERIDGE, LT.-COL. W. W. O. (1913), A Preliminary Report on the Temperature Reached in Army Biscuits during Baking, especially with reference to the Destruction of the Imported Flour-Moth *Ephestia kühniella* Zeller. *J. Roy. Army Med. Corps*, 20, 615-34.

EVELYN, J. (1641-1706), Diary.

EVERSLEY, LORD (1910), Commons, Forests and Footpaths.

FAIRCHILD, J. (1727), The City Gardener.

FARROW, E. P. (1917), On the Ecology of the Vegetation of Breckland. III., General Effects of Rabbits on the Vegetation. *J. Ecol.*, 5, 1.

FINN, F. (1905), Pekin Robins in the London Parks. *Countryside, 10,* vi. 05, 78.

FITTER, R. S. R. (1939), The Distribution of the Grey Squirrel in the London Area. *London Naturalist,* 1938, 6-19.

FITTER, R. S. R. (1940), Special Species for 1939 : British Goldfinch. *London Bird Report,* 1939, 23-25.

FITTER, R. S. R. (1941), Effect of the War on Bird Life. *Nature, 148,* 59.

FITTER, R. S. R. (1941a), Report on the Effect of the Severe Winter of 1939-40 on Bird-Life in the area within 20 miles of London. *Brit. Birds, 35,* 33-36.

FITTER, R. S. R. (1941b), Special Species for 1940 : Sand-Martin. *London Bird Report,* 1940, 16-18.

FITTER, R. S. R. (1943), The Starling Roosts of the London Area. *London Naturalist,* 1942, 3-23.

FITTER, R. S. R. (1943a), Black Redstarts in London and Middlesex in the Summer of 1942. *London Bird Report,* 1942, 17-20.

FITTER, R. S. R. (1944), Black Redstarts in England in the Summer of 1943. *Brit. Birds, 37,* 191-95.

FITTER, R. S. R. (1944a), The Black Redstart : a new British Breeding Bird. *Nature, 153,* 659.

FITTER, R. S. R. (1944b), Black Redstarts in the London Area in the Summer of 1943. *London Bird Report,* 1943, 17-19.

FITTER, R. S. R. and HOMES, R. C. (1939), Effects of the Severe Weather, December 17th-26th. *London Bird Report,* 1938, 30-33.

FITTER, R. S. R. and PARRINDER, E. R. (1944), A Check List of the Birds of the London Area. *London Bird Report,* 1943, 20-28.

FITZSTEPHEN, W. (11th cent.), Description of London : in Stow (1598).

FORBES, U. A. (1911), Sport, Ancient and Modern in *Victoria County History of Middlesex,* Vol. II.

FORD, E. B. (1940), Scientific Research in the Lepidoptera. *Ann. Eugen., 10,* 227.

GERARD, J. (1597), The Herball, or General Historie of Plantes.

GLADSTONE, HUGH S. (1919), Birds and the War.

GLEGG, W. E. (1929), A History of the Birds of Essex.

GLEGG, W. E. (1935), A History of the Birds of Middlesex.

GLEGG, W. E. (1939), Changes of Bird Life in Relation to the Increase of London. *London Bird Report,* 1938, 34-44.

GODWIN, H. (1940), Pollen Analysis and Forest History of England and Wales. *New Phyt., 39,* 370-400.

GODWIN, H. (1941), Pollen Analysis and Quaternary Geology. *Proc. Geol. Ass., 52,* 328-61.

GOMME, SIR L. (1914), London.

GRAVES, G. (1811-21), British Ornithology.

GRINLING, C. H., INGRAM, T. A., and POLKINGHORNE, B. C. (1909), A Survey and Record of Woolwich and West Kent. Woolwich.

GUENTHER, A. (1906), Reptilia and Amphibia in *Kew Bulletin*, 1906, 10-12.

GURNEY, J. H. (1921), Early Annals of Ornithology.

HALL, H. R. (1934), Unwritten History.

HAMILTON, E. (1879), The Birds of London. *Zoologist*, *3*, 273-91.

HAMPSTEAD SCIENTIFIC SOCIETY (1913), Hampstead Heath.

HARRISSON, T. H. and HOLLOM, P. A. D. (1932), The Great Crested Grebe Inquiry, 1931. *Brit. Birds*, *26*, 62-92, 102-31, 142-55, 174-95.

HARTING, J. E. (1866), The Birds of Middlesex.

HARTING, J. E. (1880), British Animals Extinct within Historic Times.

HARTING, J. E. (1886), On the Former Nesting of the Spoonbill in Middlesex. *Zoologist*, *10*, 81-88.

HARTING, J. E. (1889), The Birds of Hampstead : in Lobley (1889).

HASTINGS, A. B. (1937), Biology of Water Supply. British Museum (Natural History) Economic Series No. 7A.

HENSON, H. (1944), Entomology : some recent investigations. *Discovery*, *5*, 23-28.

HIBBERT-WARE, A. (1938), Report of the Little Owl Food Inquiry, 1936-37. (Publication of the British Trust for Ornithology.)

HICKS, H. (1892), On the Discovery of Mammoth and other Remains in Endsleigh Street and on Sections exposed in Endsleigh Gardens, Gordon Street, Gordon Square and Tavistock Square, London. *Quart. J. Geol. Soc.*, *48*, 453-68.

HINTON, M. A. C. (1931), Rats and Mice as Enemies of Mankind. British Museum (Natural History), Economic Series No. 8.

HOFLAND, T. C. (1848), British Anglers' Manual. 2nd ed.

HOLLOM, P. A. D. (1936), Great Crested Grebe Report. *London Naturalist*, 1935, 87-89.

HOME, G. (1926), Roman London.

HOME, G. (1927), Mediæval London.

HOMES, R. C. (1938), A Census of Ducks, Great Crested Grebes and Coot. *London Bird Report*, 1937, 26-28.

HORN, P. W. (1923), Notes on the Fishes of the London Docks. *London Naturalist*, 1922, 19-21.

HOWARD, A. L. (1943), The Plane Tree. *Nature*, *152*, 421.

HUDSON, W. H. (1898), Birds in London.

HUXLEY, J. S. and BEST, A. T. (1934), A Census of Water-Birds on the Highgate and Kenwood Ponds. *Brit. Birds*, 28, 122-29.

JEFFERIES, R. (1883), Nature Near London.

JOHNSON, T. (1629), Iter Plantarum Investigationis Ergo Susceptum A Decem Sociis, in Agrum Cantianum. Anno Dom. 1629. Julii 13.

Ericetum Hamstedianum. Sive Plantarum ibi crescentium observatio habita, Anno eodem 1 Augusti.

JOHNSON, T. (1632), Descriptio Itineris Plantarum Investigationis ergo Suscepti, in Agrum Cantianum Anno Dom. 1632, Et Enumeratio Plantarum in Ericeto Hamstediano locisq. vicinis Crescentium.

JOHNSON, W. (1930), Animal Life in London.

KINGSLEY, C. (1855), Glaucus.

LACK, D. (1943), The Life of the Robin.

LAVER, H. (1898), The Mammals, Reptiles and Fishes of Essex. Chelmsford.

LAVER, H. (1903), Mammalia in *Victoria County History of Essex*, Vol. I.

LEIGH, S. (1822), A New Picture of London.

LLOYD, L. (1943), Town-Planning and the Small Sewage Purification Plant. *Nature, 151*, 475-76.

LOBLEY, J. L. (1889), Hampstead Hill.

LOFTIE, W. J. (1883), A History of London.

LONDON NATURAL HISTORY SOCIETY (1900-44), Unpublished Records of Mammals, Birds, Reptiles and Amphibia in the London Area.

LONDON NATURAL HISTORY SOCIETY (1937-44), Birds of the London Area, Annual Reports for 1936-43. *London Bird Report*, 1936-43.

LOUSLEY, J. E. (1944), The Pioneer Flora of Bombed Sites in Central London. *Rep. Bot. Exch. Club*, 1941-42, 528-31.

MACPHERSON, A. H. (1928), London Reservoirs and their Influence on Bird Life. *London Naturalist*, 1927, 5-11.

MACPHERSON, A. H. (1929), A List of the Birds of Inner London, *Brit. Birds*, 22, 222-44.

MACPHERSON, A. H. (1930-40), Birds of Inner London. Annual Reports for 1929-39 in *Brit. Birds*.

MAJOR, A. F. (1920), Surrey, London and the Saxon Conquest. (Publication of the Croydon Natural History Society).

MARRIOTT, ST. J. (1925), British Woodlands as illustrated by Lessness Abbey Woods.

MARSH, G. P. (1864), Man and Nature.

MARSHALL, J. F. (1944), The Morphology and Biology of *Culex molestus* : observational notes for investigators. (Publication of the British Mosquito Control Institute, Hayling Island.)

MARSHALL, W. (1799), A Sketch of the Vale of London and an Outline of its Rural Economy.

MELVILLE, R. (1944), *Ailanthus*, Source of a Peculiar London Honey. *Nature, 154*, 640-41.

MELVILLE, R. (1945), Sources of London Honey. *Nature*, 155, 206-07.

MELVILLE, R. and SMITH, R. L. (1928), Adventive Flora of the Metropolitan Area (1) : Recent Adventives on London Rubbish. *Rep. Bot. Exch. Club*, 1927, 444-54.

MERA, A. W. (1926), Increase in Melanism in the last Half-Century. *London Naturalist*, 1925, 3-9.

MINISTRY OF AGRICULTURE AND FISHERIES (1937), Rats and How to Exterminate Them. Bulletin No. 30.

MINISTRY OF AGRICULTURE AND FISHERIES (1938), Agricultural Statistics for England and Wales for 1938.

MORGAN, D. A. T. (1939), Special Species for 1938 : Reed-Warbler. *London Bird Report*, 1938, 26-27.

MORGAN, M. T., FISHER, J. and WATSON, J. S. (1942), Preliminary Report on Rodent Control in the Area of the Port of London Health Authority.

MORGAN, M. T., FISHER, J. and WATSON, J. S. (1943), Report on Rodent Control during the year 1942 in the Area of the Port of London Health Authority.

NATIONAL SMOKE ABATEMENT SOCIETY (1938), Smoke Abatement in Greater London : Report of a Conference of London and Greater London Local Authorities on February 25, 1938.

NATIONAL SMOKE ABATEMENT SOCIETY (n.d.), The Case against Smoke : the evidence of authorities.

NATIONAL SMOKE ABATEMENT SOCIETY (n.d.a.), No Clean City.

NEWTON, J. (c. 1680), Notes as set down in his " Catalogus Plant. Angliæ " ; copied by an unknown hand in a copy of John Ray's " Catalogus Plantarum Angliæ " (1677).

NICHOLSON, C. S. (1916), The Botany of the District. *Trans. London Nat. Hist. Soc.*, 1915, 40-43.

NICHOLSON, E. M. (1926), Unpublished Survey of London Starling Roosts in 1925-26 ; quoted in Fitter (1943).

NICHOLSON, E. M. (1926a), A Bird Census of Kensington Gardens. *Discovery*, August 1926, p. 281.

NORDEN, J. (1593), Speculum Britanniae. The first parte. An Historicall . . . Description of Middlesex.

OWEN, D. J. (1927), The Port of London Yesterday and To-day.

OWENS, J. S. (1938), The Present Position and Trends of Atmospheric Pollution in Greater London : in National Smoke Abatement Society (1938), 4-12.

PAGE, W. (1923), London, its Origin and Early Development.

PARKER, E. (1941), World of Birds.

PARKINSON, J. (1640), Theatrum Botanicum ; the Theater of Plants ; or, an Herball of large extent.

PAYNE, L. G. (1942), The Royal Fern (*Osmunda regalis* L.) in Surrey. *London Naturalist*, 1941, 12-13.

PENA, P. and DE LOBEL, M. (1570), Stirpium Adversaria Nova.

PERCEVAL, P. J. S. (1909), London's Forest.

PETIVER, J. (1695), More Rare Plants Growing Wild in Middlesex : in Edmund Gibson's translation of Camden's " Britannia."

PIGOTT, T. D. (1902), London Birds and Other Sketches.

RAVEN, C. E. (1942), John Ray, Naturalist : His Life and Works. Cambridge.

RAWLENCE, D. A. (1942), Unpublished Report on the Birds and other Natural History Matters in Richmond Park in 1941.

REID, C. (1915), The Plants of the Late Glacial Deposits of the Lea Valley. *Quart. J. Geol. Soc.*, *71*, 155-63.

RITCHIE, J. (1920), The Influence of Man on Animal Life in Scotland. Cambridge.

RITCHIE, J. (1931), Beasts and Birds as Farm Pests. Edinburgh.

ROBBINS, R. W. (1916), The Flora of Epping Forest. *Trans. Lond. Nat. Hist. Soc.*, 1915, 44-48.

ROBBINS, R. W. (1927), Rabbits and Butterflies. *London Naturalist*, 1926, 37-38.

ROBBINS, R. W. (1939), Lepidoptera of a London Garden Fifty Years Ago. *London Naturalist*, 1938, 40-41.

ROWBERRY, E. C. (1934), Gulls in the London Area. *London Naturalist*, 1933, 48-58.

ROYAL COMMISSION ON HISTORICAL MONUMENTS (ENGLAND) (1928), An Inventory of the Historical Monuments in London. Vol. III., Roman London.

RUSTON, A. G. (1936), The Effects of Smoke on Vegetation. Smoke Abatement Exhibition Handbook and Guide, pp. 23-26.

SALISBURY, E. J. (1927), The Waning Flora of England. *S.E. Nat.*, 1927, 35-54.

SALISBURY, E. J. (1943), The Flora of Bombed Areas. *Nature*, *151*, 462-66.

SALMON, C. E. (1931), Flora of Surrey.

SHARPE, R. B. (1894-97), A Hand-Book to the Birds of Britain.

SHENSTONE, J. C. (1912), The Flora of London Building Sites. *J. Bot.*, *50*, 117-24.

SHIRLEY, E. P. (1867), Some Account of English Deer Parks.

SINCLAIR, R. (1937), Metropolitan Man.

SMITH, R. (1768), The Universal Directory for Destroying Rats and other kinds of Four-footed and Winged Vermin.

SPURRELL, F. J. C. (1885), Early Sites and Embankments on the Margins of the Thames Estuary. *J. Roy. Arch. Inst.*, 42, 269-302.

STOW, J. (1598), A Survey of London.

STUBBS, F. J. (1917), The London Gulls. *Trans. Lond. Nat. Hist. Soc.*, 1916, 37-41.

STUBBS, F. J. (1917a), The Mammals of the London District. *School Nature Study*, 1917.

TACITUS (1st cent. A.D.), Annals, XIV.

TALLENTS, SIR S. (1943), Man and Boy.

TANSLEY, A. G. (1939), The British Islands and their Vegetation. Cambridge.

TICEHURST, N. F. (1928), Surrey Swan Marks. *Surrey Arch. Coll.*, *38*, 34-48.

TICEHURST, N. F. (1934), The Marks Used by Swan-owners of London and Middlesex. *London Naturalist*, 1933, 67-84.

TICEHURST, N. F. (1941), The Mute Swan on the River Thames. *S.E. Nat.*, *46*, 54.

TRIMEN, H. and DYER, W. T. (1869), Flora of Middlesex.

TURNER, W. (1548), The Names of Herbes in Greke, Latin, Englishe, Duche, and Frenche, wyth the commune names that Herbaries and Apothecaries use.

TURNER, W. (1551-66), Herball. London and Cologne.

VULLIAMY, C. E. (1930), The Archæology of Middlesex and London.

WARREN, S. H. (1912), On a Late Glacial State in the Valley of the River Lea, subsequent to the Epoch of River-Drift Man. *Quart J. Geol. Soc.*, *68*, 213-51.

WARREN, S. H. (1915), Further Observations on a Late Glacial or Ponders End Stage of the Lea Valley. *Quart. J. Geol. Soc.*, *71*, 164-82.

WEBSTER, A. D. (1911), The Regent's Park and Primrose Hill.

WHITE, REV. G. (1788), The Natural History of Selborne.

WILLATTS, E. C. (1937), The Land of Britain : Part 79, Middlesex and the London Region.

WINBOLT, S. E. (1943), Britain, B.C. Harmondsworth. Penguin Books.

WITHERBY, H. F. and FITTER, R. S. R. (1942), Black Redstarts in England in the Summer of 1942. *Brit. Birds*, *36*, 132-39.

WITHERBY, H. F., JOURDAIN, F. C. R., TICEHURST, N. F. and TUCKER, B. W. (1938-41), The Handbook of British Birds.

WRIGHT, W. B. (1937), The Quaternary Ice Age.

YARRELL, W. (1843), A History of British Birds.

YARROW, I. H. H. (1941), Andrena and Nomada (Hymenoptera, Apidæ) on Hampstead Heath. *London Naturalist*, 1940, 9-13.

ZOOLOGICAL SOCIETY OF LONDON (1939), Annual Report for 1938.

ADDENDA.

CREIGHTON, C. (1891), A History of Epidemics in Britain. Cambridge.

FAIRWAY AND HAZARD, (1939), 50 miles of Golf round London.

RAMSBOTTOM, J. (1943), *Edible Fungi ;* King Penguin Books.

N.B.—The place of publication is in all cases London, except for the classics and where otherwise stated.

INDEX

Accipiter nisus, see Sparrow-hawk.
Acheulean culture, 18-19.
Acrocephalus scirpaceus, see Reed-Warbler.
Aesthetic influences, 57-58, 95-99.
Agriades corydon, see Chalkhill Blue.
Agriculture, medieval, 40.
Agricultural statistics, 185-87.
Ague, 51.
Air raids, effect of, 228 ff.
Aldenham Reservoir, 83, 158-59.
Alectoris rufa, see Partridge, Red-legged.
Algæ, 155, 162-63, 175.
Allotments, 236-37.
Amenity influences, 57-59, 95-99, 207-26.
Anas platyrhyncha, see Mallard.
Anas strepera, see Gadwall.
Anderson shelters, 134.
Angling, 94-95, 191, 204.
Anguilla anguilla, see Eel.
Ants, 113, 131.
Apothecaries, Society of, 71, 98-99.
Apus apus, see Swift.
Archaeology, 15 ff.
Ardea cinerea, see Heron.
Athene noctua, see Owl, Little.
Atmospheric pollution, 85-86, 179-84.
Aythya fuligula, see Tufted Duck.

Bacteria, 163.
Badger, 159-60, 197-98.
Barn Elms, 99.
Barn Elms Reservoirs, 166-67, 218.
Bats, 113.
Bayne, C. S., 141.
Beagles, 194.
Beavers, 50.
Bed-bug 89, 111-112.
Bees, 131, 142, 146.
Beetles, 115, 131, 152, 153, 184.
Belgravia, development of, 65.
Besant, Sir W., 28-29.
Birds, feeding of, 220.
  „   protection of, 221.
Bird Sanctuaries, 221-22.
Bird Sanctuaries in Royal Parks, Committee
  on, 140-141, 222.
Bird-catching, 81.
Bird-tables, 135, 137.
Bird-watching, 173, 238-49.
Blackberries, 192.
Black Death, 50.
Blaker, G. B., 201.

Blitzed sites, plants of, 230 ff., 265-68.
Bluebottles, 114.
Boar, wild, 91.
Bombed sites, plants of, 230 ff, 265-68.
Botanists, London, 69 ff.
Box Hill, 197.
Boyn Hill gravels, 15-18.
Brick-earth, 18.
Bramble, 192.
Branta canadensis, see Canada Goose.
Brent Reservoir, 83, 158-59, 168.
Brick-earth, 18.
British Empire Naturalists' Association, 226,
  260.
British Trust for Ornithology, 201.
Brogan, D. W., 208.
Bromley gasworks, 183.
Bronze Age, 24.
Budgerigar, 210.
Building, effects of, 30, 36-38, 63-74, 101 ff.,
  116-33.
Building materials, effects of digging for,
  148-51.
Building sites, plants of, 132-33.
Built-up areas, 116-33.
Butterflies, 131, 137, 142, 146, 184, 190.
Buxton, E. N., 197, 198.

Canada Goose, 218.
Canals, influence of, 83, 155-59.
Capella gallinago, see Snipe, Common.
Capreolus capreolus, see Roe Deer.
Carduelis carduelis, see Goldfinch.
Carpinus betulus, see Hornbeam.
Carter, J. S., 138.
Castell, C. P., 234.
Cats, 58, 209-210.
Cattle, 78, 187.
Cedar of Lebanon, 96.
Cedrus libani, see Cedar of Lebanon.
Cervus elaphus, see Deer, Red.
Chalk, 9.
Chalkhill Blue, 190.
Charadrius dubius, see Plover, Little Ringed.
Chelsea Physic Garden, 73, 79, 97, 99, 225.
Charcoal burning, 84-85.
Charles I, 92.
Chellean culture, 16-18.
Chenopodium, see Goosefoot.
Chiswick Eyot, 190.
Cimex lectularius, see Bed-bug.
City Corporation, 222.
Climate, prehistoric, 16 ff.

Clothing, effects of utilisation for, 18, 25, 49, 50.
Clusius, C., 69, 86, **98.**
Cockroaches, 89, 113.
Collenette, C. L., 136, 198.
Collinge, W. E., 200.
Coltsfoot, 232.
Columba palumbus, see Woodpigeon.
Commerce, influence of, 59-60, 88-89, 152-161.
Commons, enclosure of, 80.
Conifers, National Collection of, 182.
Convallaria majalis, see Lily of the Valley.
Cormorant, 219.
Cornish, C. J., 166, 171-73, 191, 203.
Corvus corax, see Raven.
Corvus corone, see Crow, Carrion,
Corvus frugilegus, see Rook.
Corvus monedula, see Jackdaw.
Cobbett, W., 61, 76, 99.
Coturnix coturnix, see Quail.
Crake, Spotted, 69.
Cricket, 113.
Cro-Magnon man, 122-23.
Crow, Carrion, 118.
Cubitt, Thomas, 65.
Cuckoo, 109-110, 190.
Cuculus canorus, see Cuckoo.
Cultivation, effect of, 23, 25, 38-43, 75-80, 185-90.
Curtis, William, 68, 72-73, 79, 82.
Cygnus olor, see Swan, Mute.

Dabchick, 142.
Dagenham, 176.
Dama dama, see Deer, Fallow.
Danes, ravages of, 34.
Dark Ages, 32-34.
Dartford Warbler, 108-109.
Deer, 53-56.
   „   Fallow, 91, 197, 213-14.
   „   Red, 91, 197, 213-14.
   „   Roe, 197.
Defoe, D., 76.
De L'Ecluse, C., 69, 86, 98.
Dent, Geoffrey, 197, 198.
Deptford Pink, 71.
Dianthus armeria, see Deptford Pink.
Divers, 165.
Dixon, C., 230.
Docks, influence of, 154-55.
Dogs, 54, 209.
Dogs, Isle of, 64.
Domesday Book, 39-40, 42, 49-50.
Domestication, effects of, 23-25, 38-43.
Donnelly, R. P., 183.
Dragonflies, 131.
Dreissensia polymorpha, see Zebra Mussel.

Duck decoys, 93.
Ducks, 158, 165, 168, 173.
Dunghills, plants growing on, 82, **86.**

Eales-White, Major J. C., 136.
Eels, 46, 81, 94, 164, 171, 191.
Elephants, 32.
Elephas primigenius, see Mammoth.
Elstree Reservoir, 83, 158-59.
Emys orbicularis, see Tortoise, Water.
Endsleigh St., excavation in, 19.
Epilobium angustifolium, see Rose-bay willow-herb.
Epping Forest, 53-55, 91, 182, 192, 194, 196, 197, 198, 216, 222.
Erigeron canadensis, see Fleabane, Canadian.
Erinaceus europaeus, see Hedgehog.
Erithacus rubecula, see Robin.
Escapes, 140, 210 ff., 238.
Evelyn, John, 85, 92, 95, 180.

Falco peregrinus, see **Peregrine.**
Falco subbuteo, see Hobby.
Falco tinnunculus, see Kestrel.
Famines, 46.
Ferns, 224.
Finn, F., 211.
Fire of 1666, 18, 34, 65, 89, 132.
Fires in London, 34.
Firewood, cutting of, 49, 84.
Fish, 81, 94, 154-55, 171-73, 235.
Fish-weirs, 46-47.
Fisher, James, 130.
Fishing, 94-95, 191, 204.
Fitzstephen, W., 35-36, 52.
Five Fields, 65.
Fleabane, Canadian, 232.
Fleas, 111.
Fleet river, 37.
Flies, 114, 142, 175.
Flood Plain gravels, 20-22.
Flounder, 81, 94.
Fogs, influence of, 180.
Ford, E. B., 182.
Forest, definition of, 53.
Forest laws, 53-54.
Food, effects of utilisation for, 16-17, 25, 44-49, 80-82, 190-92.
Fossils, 9.
Fox, 193-95.
Foxhounds, 193-94.
Frog, Edible, 211.
Fungi, 115, 192.

Gadwall, 218.
Galinsoga parviflora, 233.
Gallinula chloropus, see Moorhen.
Game, protection of, 93, 199-203.

Gardens, 42, 58, 95, 97, 134-40, 224.
Gasworks, flora and fauna of, 183-84.
Geology, 8 ff.
Georgian London, 61 ff.
Gerard, J., 69, 70, 77, 97.
Gesner, 58 n., 94.
Ginkgo biloba, see Maidenhair tree.
Glasshouses, 182.
Goldfinch, 188-89, 221.
Golf courses, 204-05.
Gomme, Sir L., 52.
Goosefoots, 82, 86.
Grand Junction Canal, 83.
Grand Union Canal, 83.
Grasses, 143.
Grasshopper-warbler, 108.
Gravel pits, flooded, 21, 148-51.
Gravel terraces, 14-22.
Graves, G., 68.
Gray's Inn, 106.
Grebe, Great Crested, 149-221.
  „ Little, 142.
Grebes, 165.
Ground Game Act, 196.
Groundsel, 232.
  „ Sticky, 232.
Gryllulus domesticus, see Cricket.
Guichard, K. M., 146.
Gull, Black-headed, 176-78.
  „ Common, 220.
Gulls, 174, 176-78, 205.
Gurney, J. H., 35.

Hainault Forest, 222.
Hampstead Heath, 70-72, 143-47, 148, 192, 194, 198, 204, 216, 226, 244-48.
Hampton Court Maze, 24.
Harding, J. Rudge, 211.
Hare, 196.
Harriers (birds), 68.
Harriers (hounds), 194.
Harting, J. E., 80, 83, 144, 168, 176, 191, 210.
Hastings, Dr. A., 162.
Hawking, 56-57, 90.
Haymaking, 187.
Hedgehog, 134.
Henry VIII, 89.
Herb-gathering, 42, 48, 82, 192.
Heron, 230.
High Terrace gravels, 15-18.
Highgate Ponds, 144, 146, 219.
Highgate Wood, 226.
Hill, Sir Arthur, 182.
Hinton, M. A. C., 59, 112.
Hobby, 200.
Home, G., 47.
Honey, 142-43.

Horn, P. W., 154-55, 211.
Hornbeam, 24.
Horses, 187, 213.
Hounds, 193-94.
Hudson, W. H., 105, 107, 118, 129, 142, 177, 204, 209, 220, 221.
Hunting, effects of, 52-57, 89-95, 193-98.
Huxley, J. S., 144.
Hyde Park, 91-92, 196, 118, 142, 195.

Ice Age, 15 ff., 44.
Indoor fauna, 111-15.
Introductions of animals and plants, 210 ff.
Iron age, 24.

Jackdaw, 106-07.
Jefferies, Richard, 160, 204 n.
Johnson, Thomas, 68, 69, 71-72, 99, 244.
Johnson, W., 113, 131, 173, 183, 199, 204.

Kensington Gardens, 105, 126, 141, 142, 195, 215.
Ken Wood, *see* Hampstead Heath.
Kestrel, 118, 183, 200.
Kew Gardens, 97, 181, 191, 215, 217, 224, 225.
Kingsley, Charles, 203.
Kite, 51-52, 86-87.

Lack, David, 137.
Land Utilisation Survey, 186.
Larus canus, see Gull, Common.
Larus ridibundus, see Gull, Black-headed.
Laver, H., 196.
Lawns, 139.
Lea, gravel terraces of, 16-22.
Lepisma, see Silver-fish.
Lepidoptera, *see* Butterflies, Moths.
Lepus europaeus, see Hare.
Lice, 111.
Lily of the Valley, 70, 72, 97.
Linnæus, C., 99.
Lobel, M. de, 69, 71.
Locustella naevia, see Grasshopper-warbler.
Londinium, 27-33.
London, expansion of, 35, 61 ff., 101-04.
  „ foundation of, 27.
London area, boundary of, 6-7.
  „ „ definition of, 6.
London Clay, 11, 14, 21, 263.
London County, boundary of, 6.
London County Council, 172, 222.
London Natural History Society, 6, 226, 259.
London Rocket, 132, 133, 231.
Lousley, J. E., 231, 233.
Low, G. C., 219.
Low Terrace gravels, 20-22.
Luscinia megarhyncha, see Nightingale.
Lutra lutra, see Otter.

Macpherson, A. Holte, 109.
Maianthemum bifolium, see May Lily.
Mallard, 173, 236.
Maidenhair-tree, 97.
Mammoth, 16 ff.
Mann, E., 195.
Man's effect on balance of Nature, 16-17, 24-25.
Market gardening, 21, 77, 79.
Martes martes, see Marten.
Marten, 199.
Martin, Sand, 14, 150-51.
Matheson, Colin, 209.
May Day, 58, 98.
May lily, 147.
Melanism, industrial, 182.
Meles meles, see Badger.
Melopsittacus undulatus, see Budgerigar.
Melville, R., 142, 176.
Mentha pulegium, see Pennyroyal.
Mesolithic Culture, 22-23.
Metropolitan Board of Works, 62, 171.
Metropolitan Water Board, 83, 162, 165.
Middle Ages, 34-60.
Middle Terrace gravels, 18-19.
Middlesex, Forest of, 23, 30.
    „   lost plants of, 254-55.
Milvus milvus, see Kite.
Mites, 152-53.
Mollusca, 89, 164.
Moorfields, 32, 51.
Moorhen, 142.
Morgan, M. T., 153.
Mosses, 133.
Moths, 138-39, 142.
    „   Clothes, 114.
    „   House, 114-15.
Motor cars, influence of, 159-60.
Moulds, 115.
Mouse, House, 112-13, 153-54.
Mousterian culture, 19-20.
Mulberry tree, 79.
Mustela erminea, see Stoat.
Mustela nivalis, see Weasel.
Mustela putorius, see Polecat.
Museums, 261-62.
Mus musculus, see Mouse, House.

National Trust, 222.
Natural history societies, 226, 259-61.
Nature, appreciation of, 98-99, 207-09.
Neanderthal man, 19-20.
Neolithic culture, 23-24.
New River, 82.
Nicholson, C., 226.
Nicholson, E. M., 121, 126, 141.
Nightingale, 109, 210.
Norden, J., 75.

Normans, introductions by, 45, 57.
Nuthatch, 107.

Oenanthe oenanthe, see, Wheatear.
Orchards, 42, 187.
Orgyia antiqua, see Vapourer.
Oryctolagus cuniculus, see Rabbit.
Osiers, 189.
Otter, 160, 195-96.
Otterhounds, 194.
Owl, Barn, 201-02.
    „   Brown, 202.
    „   Little, 201.
    „   Tawny, 202.

Palaeolithic culture, 16-22.
Panurus biarmicus, see Tit, Bearded.
Parietaria officinalis, see Pellitory.
Paris, Matthew, 43.
Parker, Eric, 130.
Parkinson, J., 69, 71.
Parks, 56, 95-96, 140-47, 222 ff.
Parliament Hill Fields, *see* Hampstead Heath.
Parrots, 58.
Partridge, French, 93, 198-99.
Partridge, Red-legged, 93, 199.
Passer domesticus, see Sparrow, House.
Pelican, 95, 219.
Pellitory, 133.
Pennant, T., 87.
Pennyroyal, 70.
Peregrine Falcon, 118.
Pethen, R. W., 195.
Petiver, J., 69.
Pets, 58, 209-11.
Phalacrocorax carbo, see Cormorant.
Phasianus colchicus, see Pheasant.
Pheasant, 57, 198.
Phoenicurus ochrurus, see Redstart, Black.
Phragmites communis, see Reed, Common.
Pigeons, 59, 119-20, 229.
Pigott, T. D., 129, 131, 204.
Pigs, 40, 51, 87-88.
Pine-Marten, 199.
Plagues, 50.
Plane-trees, 86.
Platalea leucorodia, see Spoonbill.
Platichthys flesus, see Flounder.
Playing fields, 204-05.
Plover, Little Ringed, 150.
Podiceps cristatus, see Grebe, Great Crested.
Podiceps ruficollis, see Dabchick.
Polecat, 199.
Ponds, effects of draining, 235.
Population, 29, 31, 62-63, 101.
Porzana porzana, see Crake, Spotted.
Prices, medieval, 47-48.
Prunella modularis, see Sparrow, Hedge.

Quail, 199.

Rabbit, 44-46, 183, 190-91.
Ragwort, Oxford, 232.
Railway embankments, plants on, 86, 160-61.
Railways, influence of, 159-60.
Rana esculenta, see Frog, Edible.
Rat, Black, 59-60, 112, 153-54.
„ Brown, 88-89, 112, 153-54, 176.
Rattus norvegicus, see Rat, Brown.
Rattus rattus, see Rat, Black.
Raven, 51, 87.
Rawlence, D. A., 235.
Ray, John, 78 *n.*, 82, 96, 231 *n.*
Redstart, Black, 120-26.
Reed, Common, 108.
Reed-warbler, 108.
Refuse disposal, effects of, 50-52, 86-88, 114, 171-78.
Regent's Canal, 83, 155, 158.
Regent's Park, 143, 214, 225.
Reid, Clement, 21.
Reservoirs, influence of, 83, 158-59, 165-70.
Richmond Park, 198, 214, 221, 230, 235.
Ring-dove, 141-42.
Riparia riparia, see Martin, Sand.
Ritchie, Prof. J., 88, 211.
Robbins, R. W., 138, 190-91, 224.
Robin, 108, 136, 137.
Rocque's map, 77.
Roe-deer, 197.
Romans, influence of, 27-33.
Rook, 105-06.
Rose-bay willow-herb, 73, 132, 231.
Rowberry, E. C., 205.
Royal Society for the Protection of Birds, 201.
Rubus fruticosus, see Bramble.
Ruislip Reservoir, 83, 158-59.
Ruston, A. G., 181.

Safety of man and stock, effects of ensuring, 30, 43-44.
Salisbury, E. J., 23, 32, 110, 231-34, 265.
Salmon, 82, 94.
Salmo salar, see Salmon.
Saxons, London, 34 ff.
Scavengers, animals as, 50-52, 86-88, 176-78.
Scirpus triqueter, 69.
Sciurus cinereus, see Squirrel, Grey.
Sciurus vulgaris, see Squirrel, Red.
Scolopax rusticola, see Woodcock.
Selborne Society, 222.
Senecio squalidus, see Ragwort, Oxford.
Senecio viscosus, see Groundsel, Sticky.
Senecio vulgaris, see Groundsel.
Serpentine, 96.

Sewage farms, influence of, 174-75.
Sharpe, Bowdler, 81.
Shenstone, J. C., 132.
Shooting, 93, 199-203.
Silver-fish, 115.
Sitta europaea, see Nuthatch.
Sisymbrium irio, see London Rocket.
Smoke, influence of, 85-86, 179-84.
Snails, 89, 164, 224.
Snipe, Common, 65.
Sparrow, Hedge, 136.
„ House, 118-19.
Sparrow-hawk, 200-201.
Spiders, 113-14.
Sponges, 164.
Spoonbill, 80.
Sport, effects of, 52-57, 89-95, 193-205.
Spring Gardens, excavation in, 21-22.
Squirrel, Grey, 214-16.
„ Red, 216.
Staines Reservoirs, 169.
Stamp, Dudley, 236 *n.*
Starling, 126-31, 174.
Static water tanks, 236.
Stepney, manor of, 39.
St. James's Park, 92, 95-96, 129, 142, 177, 210.
Stoat, 201.
Stone Age, 16-23.
Stow, J., 47, 51, 76-77, 81, 82, 87.
Strix aluco, see Owl, Tawny.
Stuart London, 61 ff.
Sturnus vulgaris, see Starling.
Sunshine, effect of smoke on, 179-80.
Sus scrofa, see Boar, Wild.
Swan, Mute, 41-42, 80, 219-20.
Swan-keeping, 41-42.
Swan-upping, 41-42, 219.
Swift, 118.
Sylvia undata, see Dartford Warbler.
Syon Park, 69-70.

Taplow gravels, 18-19.
Thames, embankment of, 28, 43.
„ gravel terraces of, 14-22.
„ pollution of, 94, 171.
„ towpath, 173, 238.
Thrush, Song, 136.
Tit, Bearded, 68.
Tortoise, Water, 211.
Tower of London, 58-59, 96.
Trade, influence of, 59-60, 88-89, 152-61.
Traffic, influence of, 159-60.
Trimen & Dyer, 68, 71, 73, 231.
Tring Reservoirs, 150, 158.
Tudor London, 61 ff.
Tufted Duck, 219.
Turdus ericetorum, see Thrush, Song.

Turner, W., 69, 94.
Turner's Wood, 146.
Tussilago farfara, see Coltsfoot.
Tyto alba, see Owl, Barn.

Vapourer moth, 131.
Vineyards, 39, 42-43.
Vulpes vulpes, see Fox.

Waders, 169, 174.
Walbrook, 31, 37.
Warehouses, fauna of, 152-53.
Warren, S H., 22.
Walls, flora of, 71, 133.
Walthamstow Reservoirs, 166, 218, 230.
Walton, Isaac, 94.
War, effect of, 48 *n.*, 228-39.
Waste ground, plants of, 132-33, 175-76, 230 ff., 265-68.
Water supply, effects of, 82-84, 162-70.
Watercress, 190.
Waterfowl, 149-50, 158-59, 165-68, 173, 217-18.
Waterhen, 142.
Waterworks, flora and fauna of, 162-65
Weasel, 201.

Weather, severe, effects of, 109, 220.
Webster, A. D., 196.
Weeds, 139.
Welsh Harp, 83.
Wembley, 120-21.
Wen, the, 61, 76.
Westbourne river, 66, 96.
Westminster, manor of, 39.
Wheatear, 234.
White, Gilbert, 74.
Whittington, Dick, 58.
Willingale, Thomas, 222.
Windsor, 7, 9, 91.
Wolves, 43.
Woodcock, 65.
Woodlands, destruction of, 25, 31, 38-39 49, 77-78, 84-85, 237.
Woodpeckers, 237.
Woodpigeons, 141-42.
Worms, 112, 164, 175.

Yarrell, W., 107.
Yarrow, I. H. H., 146.

Zebra mussel, 164.
Zoo, 58-59, 96, 150, 212-13.

Photogravure Plate (I) "Aerial View of London and the Thames Estuary," and Plate (II) "Central London from a tall building near St. Paul's are reproduced by courtesy of *The Times*.